International Conference on Bioinformatics & Computational Biology 2016 (BIOCOMP 2016)

Held at WORLDCOMP'16

Las Vegas, Nevada, USA
25 – 28 July 2016

Editors:

Hamid R. Arabnia
Fernando G. Tinetti
Quoc-Nam Tran
Mary Yang
George Jandieri
Ashu M. G. Solo

ISBN: 978-1-5108-3414-9

Printed from e-media with permission by:

Curran Associates, Inc.
57 Morehouse Lane
Red Hook, NY 12571

Some format issues inherent in the e-media version may also appear in this print version.

Printed by Curran Associates, Inc. (2017)

For permission requests, please contact CSREA Press
at the address below.

CSREA Press
ATTN: Hamid R. Arabnia
1060 Long Creek Drive
Watkinsville, GA 30677
USA

Phone: (706) 542-3480

hra@cs.uga.edu

Additional copies of this publication are available from:

Curran Associates, Inc.
57 Morehouse Lane
Red Hook, NY 12571 USA
Phone: 845-758-0400
Fax: 845-758-2633
Email: curran@proceedings.com
Web: www.proceedings.com

PROCEEDINGS OF
THE 2016 INTERNATIONAL CONFERENCE ON BIOINFORMATICS & COMPUTATIONAL BIOLOGY

BIOCOMP 2016

Editors

Hamid R. Arabnia, Fernando G. Tinetti
Quoc-Nam Tran, Mary Yang

Associate Editors

George Jandieri, Ashu M. G. Solo

WORLDCOMP'16
July 25-28, 2016
Las Vegas Nevada, USA
www.worldcomp.org

This volume contains papers presented at The 2016 International Conference on Bioinformatics & Computational Biology (BIOCOMP'16). Their inclusion in this publication does not necessarily constitute endorsements by editors or by the publisher.

CSREA Press
U. S. A.

Foreword

It gives us great pleasure to introduce this collection of papers to be presented at the 2016 International Conference on Bioinformatics & Computational Biology (BIOCOMP'16), July 25-28, 2016, at Monte Carlo Resort, Las Vegas, USA.

An important mission of the World Congress in Computer Science, Computer Engineering, and Applied Computing (a federated congress to which this conference is affiliated with) includes *"Providing a unique platform for a diverse community of constituents composed of scholars, researchers, developers, educators, and practitioners. The Congress makes concerted effort to reach out to participants affiliated with diverse entities (such as: universities, institutions, corporations, government agencies, and research centers/labs) from all over the world. The congress also attempts to connect participants from institutions that have* ***teaching*** *as their main mission with those who are affiliated with institutions that have* ***research*** *as their main mission. The congress uses a quota system to achieve its institution and geography diversity objectives."* By any definition of diversity, this congress is among the most diverse scientific meeting in USA. We are proud to report that this federated congress has authors and participants from 74 different nations representing variety of personal and scientific experiences that arise from differences in culture and values. As can be seen (see below), the program committee of this conference as well as the program committee of all other tracks of the federated congress are as diverse as its authors and participants.

The program committee would like to thank all those who submitted papers for consideration. About 50% of the submissions were from outside the United States. Each submitted paper was peer-reviewed by two experts in the field for originality, significance, clarity, impact, and soundness. In cases of contradictory recommendations, a member of the conference program committee was charged to make the final decision; often, this involved seeking help from additional referees. In addition, papers whose authors included a member of the conference program committee were evaluated using the double-blinded review process. One exception to the above evaluation process was for papers that were submitted directly to chairs/organizers of pre-approved sessions/workshops; in these cases, the chairs/organizers were responsible for the evaluation of such submissions. The overall paper acceptance rate for regular papers was 27%; 16% of the remaining papers were accepted as poster papers (at the time of this writing, we had not yet received the acceptance rate for a couple of individual tracks.)

We are very grateful to the many colleagues who offered their services in organizing the conference. In particular, we would like to thank the members of Program Committee of BIOCOMP'16, members of the congress Steering Committee, and members of the committees of federated congress tracks that have topics within the scope of BIOCOMP. Many individuals listed below, will be requested after the conference to provide their expertise and services for selecting papers for publication (extended versions) in journal special issues as well as for publication in a set of research books (to be prepared for publishers including: Springer, Elsevier, BMC journals, and others).

- *Prof. Afrand Agah (ABDA'16); Department of Computer Science, West Chester University of Pennsylvania, West Chester, PA, USA*
- *Prof. Abbas M. Al-Bakry (Congress Steering Committee); University President, University of IT and Communications, Baghdad, Iraq*
- *Prof. Nizar Al-Holou (Congress Steering Committee); Professor and Chair, Electrical and Computer Engineering Department; Vice Chair, IEEE/SEM-Computer Chapter; University of Detroit Mercy, Detroit, Michigan, USA*
- *Prof. Hamid R. Arabnia (Congress Steering Committee & Coordinator); Graduate Program Director (PhD, MS, MAMS); The University of Georgia, USA; Editor-in-Chief, Journal of Supercomputing (Springer); Editor-in-Chief, Transactions of Computational Science & Computational Intelligence (Springer); Fellow, Center of Excellence in Terrorism, Resilience, Intelligence & Organized Crime Research (CENTRIC).*
- *Prof. P. Balasubramanian (PDPTA'16); School of Computer Engineering, Nanyang Technological University, Singapore*
- *Prof. Juan Jose Martinez Castillo; Director, The Acantelys Alan Turing Nikola Tesla Research Group and GIPEB, Universidad Nacional Abierta, Venezuela*

(CCISA); Department of Information Management, Central Police University, Taoyuan, Taiwan; Guest Ed., IEEE Journal on Selected Areas in Communications.

- *Prof. Mary Yang (Co-Editor); Director, Mid-South Bioinformatics Center and Joint Bioinformatics Ph.D. Program, Medical Sciences and George W. Donaghey College of Engineering and Information Technology, University of Arkansas, USA*
- *Prof. Jane You (Congress Steering Committee & Vice-Chair of IPCV'16); Associate Head, Department of Computing, The Hong Kong Polytechnic University, Kowloon, Hong Kong*

We would like to extend our appreciation to the referees, the members of the program committees of individual sessions, tracks, and workshops; their names do not appear in this document; they are listed on the web sites of individual tracks.

As Sponsors-at-large, partners, and/or organizers each of the followings (separated by semicolons) provided help for at least one track of the Congress: Computer Science Research, Education, and Applications Press (CSREA); US Chapter of World Academy of Science (http://www.worldcomp.org/) ; American Council on Science & Education & Federated Research Council (http://www.americancse.org/); HoIP, Health Without Boundaries, Healthcare over Internet Protocol, UK (http://www.hoip.eu); HoIP Telecom, UK (http://www.hoip-telecom.co.uk); and WABT, Human Health Medicine, UNESCO NGOs, Paris, France (http://www.thewabt.com/). In addition, a number of university faculty members and their staff (names appear on the cover of the set of proceedings), several publishers of computer science and computer engineering books and journals, chapters and/or task forces of computer science associations/organizations from 3 regions, and developers of high-performance machines and systems provided significant help in organizing the conference as well as providing some resources. We are grateful to them all.

We express our gratitude to keynote, invited, and individual conference/tracks and tutorial speakers - the list of speakers appears on the conference web site. We would also like to thank the followings: UCMSS (Universal Conference Management Systems & Support, California, USA) for managing all aspects of the conference; Dr. Tim Field of APC for coordinating and managing the printing of the proceedings; and the staff of Monte Carlo Resort (Convention department) in Las Vegas for the professional service they provided. Last but not least, we would like to thank the Co-Editors and Associate Co-Editors of BIOCOMP'16: Prof. Hamid R. Arabnia, Prof. George Jandieri, Prof. Fernando G. Tinetti, Prof. Quoc-Nam Tran, Prof. Mary Yang, and Ashu M. G. Solo.

We present the proceedings of BIOCOMP'16.

Steering Committee, 2016
http://www.worldcomp.org/

Contents

SESSION: PROTEIN CLASSIFICATION, STRUCTURE PREDICTION, AND COMPUTATIONAL STRUCTURAL BIOLOGY

Application of msTALI in ATPase Active Site Identification 3
Devaun McFarland, Caroline Bullock, Benjamin Mueller, Homayoun Valafar

Protein Structure-Function Analysis with Self-Organizing Maps 10
Seonjoo Lim, Stephen Jaegle, Lutz Hamel

A Support Vector Machine Based Model for Predicting Heparin-Binding Proteins Using XB Patterns as Features 17
Joseph Sirrianni, Zhichun Xiao, Wing Ning Li

Phosphatidylserine Torus as a Macromolecular Scaffold for the GABAa Receptor 22
Sara Reader, Julie Mustard, Norbert Seidler

SESSION: GENE EXPRESSION, REGULATORY NETWORKS, MICROARRAY, SEQUENCING, ALIGNMENT, AND RELATED STUDIES

An Anomaly Detection Algorithm for Identifying Alien Gene Clusters in Microbial Genomes 31
Dongsheng Che, Sai Manikonda, Zuqing Li, Bernard Chen

PGAR: ASD Candidate Gene Prioritization System Using Expression Patterns 38
Steven Cogill, Liangjiang Wang

Information-theoretic Interestingness Measures for Cross-Ontology Data Mining 44
Prashanti Manda, Fiona McCarthy, Bindu Nanduri, Hui Wang, Susan M. Bridges

Inferring Gene Network from Gene Expression Data Using Dynamic Bayesian Network with Optimization Algorithm and Scoring Metric Approaches 50
Muhammad Mahfuz Zainuddin, Mohd Saberi Mohamad, Lian En Chai, Zuraini Ali Shah, Weng Howe Chan, Safaai Deris, Hussah AlEisa, Saad Subair

Increasing Efficiency of Microarray Analysis by PCA and Machine Learning Methods 56
Jing Sun, Kalpdrum Passi, Chakresh Jain

SESSION: SIGNAL AND DATA PROCESSING, CLUSTERING METHODS, IMAGING SCIENCE, AND DATA QUALITY ENHANCEMENT

Large Scale SVS Images Stitching for Osteosarcoma Identification 67
Bogdan Armaselu, Harish Arunachalam, Ovidiu Daescu, John-Paul Bach, Kevin Cederberg, Sam Glick, Dinesh Rakheja, Anita Sengupta, Stephen Skapek, Patrick Leavey

Optimizing the Detection of Events of Interest in Serial Data 74
Robert Warner

Detecting Frequency from Randomly Sampled Data Implementation of random sampling in BRATUMASS 81
Luxi Li, Yizhou Yao, Meng Yao, Erik D. Goodman, E. John R. Deller

Research on Approach for Classification of Within Imbalanced Data Sets 85
Chunkai Zhang, Jiayao Jiang, Fengxing Shi

A Proposed Warped Choi Williams Time Frequency Distribution Applied to Doppler Blood Flow Measurement 90
F. Garcia-Nocetti, J. Solano, E. Rubio

SESSION: SYSTEMS BIOLOGY AND RNA SEQUENCE DATA PROCESSING + HIGH PERFORMANCE COMPUTING

Stage-specific Transcriptome of the Malaria Parasite in the Red Blood Cell Cycle 97
Hao Zhang, Timothy Lilburn, Hong Cai, Yufeng Wang

Acceleration of Computational Fluid Dynamics Analysis by using Multiple GPUs 103
Hyungdo Lee, Bongjae Kim, Kyounghak Lee, Hyedong Jung

RNAseq Analysis of C. elegans Infected by Orsay Virus Suggests Evidence of a New RNAi or Stress Related Pathway 109
Jessica Ngo, Johanshah Ashkani, Frederic Pio

SESSION: NOVEL STUDIES

Recurrent Breast Cancer Treatment via Rapid Pain-Guided Experimentation 117
Jane Eyre, Steve Richfield

SESSION: POSTER PAPERS

A Gain Compensation Algorithm for Hearing Aid using the Voice Activity Detection 127
Sang-Kyun Kim, Sang-Ick Kang, Young-Jin Park, Jang-Woo Kwon, Sangmin Lee

Broadening the Scope of Computer Science Education: Introducing and Incorporating Biodiversity Informatics education to a Computer Science Curriculum 129
Nazli Hardy, Christopher Hardy

Predicting the Function of Hypothetical Protein PANDA_003700 using Computational Analysis Methods 130
Cameron Bixby, Padmanabhan Mahadevan

SESSION: LATE BREAKING PAPERS

An Ecological-Based Genetic Investigation Using Association Analysis Approach 135
Ray R. Hashemi, Azita A. Bahrami, Jeffery A. Young, Aaron Schrey

SESSION

PROTEIN CLASSIFICATION, STRUCTURE PREDICTION, AND COMPUTATIONAL STRUCTURAL BIOLOGY

Chair(s)

TBA

Application of msTALI in ATPase Active Site Identification

Devaun McFarland[1], Caroline Bullock[2], Benjamin Mueller[2], and Homayoun Valafar[1*]

[1]Department of Computer Science and Engineering, University of South Carolina, Columbia, SC 29208, USA

[2]Department of Biomedical Engineering, University of South Carolina, Columbia, SC 29208, USA

[*] **Corresponding Author Email: homayoun@cec.sc.edu Phone: 1 803 777 2404 Fax: 1 803 777 3767**

Mailing Address: Swearingen Engineering Center, Department of Computer Science and Engineering, University of South Carolina, Columbia, SC 29208, USA

Abstract - *Proteins constitute an important class of biomolecules that are analogous to workers in a factory. Performing various activities in a biological cell, proteins provide structural support and facilitate important biochemical reactions. Active sites are areas where reactions and bindings events take place and therefore, they describe a protein's function. Function of a protein depends on factors such as the structure of its active site and its relationship with the ligand/s. Successful identification of active sites is significant for understanding molecular basis for diseases, assisting in drug design, the study of targeting mutants, and for functional annotation of unknown proteins. Better understanding of active sites is beneficial in protein design and engineering. This research outlines msTALI as a suitable strategy for addressing active site identification and compares the results of msTALI to other existing methods. We report successful identification of a motif characteristic to ATPase activity using msTALI.*

Keywords: msTALI, conserved regions, ATPase, active sites, protein ligands, corresponding residues

1 Introduction

Active site identification is salient to better understanding of the mechanism by which proteins operate. Structural information is useful for meaningful investigation of active sites. Active sites are the areas where bindings occur and by this, are synonymous, and describe the function of a protein. Proteins with a similar function maintain a common structural regions, the recovery of which requires the sequence-structure-function relationships remain intact. Still, the sequence-structure-function relationship is not always easy to describe, classify, and in some cases even recognize. The functionality of a protein relies on conformity of active site location, size, and its relationship with the ligands it binds. Successfully identifying active sites is significant for understanding molecular basis for diseases, assisting in drug design, the study of targeting mutants, and for functional annotation of unknown proteins; it's beneficial in protein design and engineering as well.

Many computational methods seek to solve this problem effectively. Surface of proteins are frequently irregular and the descriptions of cavities make such studies nontrivial [1]. Consequently, a common practice is the exploration of docking, where surface accessible regions are coupled with ligands specific to the interaction. Literature suggests that geometrical methods are at the forefront for addressing active site identification, and that often an integration of chemical components is necessary. One of the first geometric approaches was grid based [2]. Still, depending on orientation within a three-dimensional space, initial grid methods had shortcomings. LIGSITE expanded this construct by elaborating rankings of cavities based off of relevant binding sizes and expansion of its algorithmic processes in breadth [1, 3]. Another approach described is SURFNET; it depicts the use of spheres amongst atomic pairs to generate groupings from within a protein. Additionally surface spheres invoke a framework for identifying the largest cavities for successful binding [4]. All of which are important for the development of binding strategies. Spheres and convex hull representations all support the consensus that size parameters of cavities are relevant [5]. Amongst other computational methods are hashing techniques and graph theoretic methods [6]. Neural Networks (NN) have also been used for comparing structural similarities. The construct of using NN training is premised by the idea that proteins with similar structures have similar functionalities [7]. Yet the function to structure relationship isn't direct. Proteins perform many functions; the NN is then used for scoring procedures that classify catalytic regions. Further, either training is required, or there are not exact localized descriptions about the proteins. In the latter case, fuzzy functional forms (FFFs) have been adapted for predictions too [8][9]. Last there are web servers available that seek to identify active sites, and protein documenting directories such as Pfam that incorporate markup information for active sites [10][11][12].

Existing computational approaches to active site identification fall short of ideal by failing to include some critical information such as: global and local structure, amino acid position, and local biochemical properties. Numerous and valuable advances have aimed to address some of the existing shortcomings individually, but the Multiple Structure Torsion Angle Alignment (msTALI) approach addresses many of the shortcomings concurrently [13]. Here, we propose a novel utility of the msTALI approach addressing the active site identification problem. The msTALI's alignments generate competitive results and outperform software that are scored and rest on the use of backbone RMSD values for structural alignment [6, 14]. Our methodology takes advantage of the existing engine by performing superior alignments on proteins that are documented to perform the same function, all while detecting dynamic confirmations; this becomes key when addressing proteins classified with similar function, that bind flexible ligands, and are non-homologous [15].

2 Background and methods

The general overview of our approach is to utilize msTALI in performing multiple structure alignment of 19 proteins and compare the results to previous work. These proteins have biologically been confirmed to exhibit ATPase activity and previous structure based analyses have been reported in recent publications. The following sections provide an overview of the background materials and methods.

2.1 Target proteins

Our structure-based identification of active sites relied on analysis of 19 proteins with confirmed ATPase activity. The selection of these proteins was based on three contributing factors: existence of previous work for comparison purposes, structural diversity, and complexity of the problem. These 19 proteins were subject of a previous analysis using Continuous Optimization (CO)[6] and Molecular Local Surface Comparison (MolLoc) [14].

Analysis of ATPase proteins is compelling in two additional aspects; protein dissimilarity, and the structural flexibility of ATP as a substrate. The diversity of proteins that exhibit ATPase activity renders this problem particularly challenging and therefore meaningful to address. Table 1 lists the 19 target proteins with some of their binding properties. Column two describes the organisms associated with each protein and highlights a plethora of organisms, ranging from wild boar to human flu. Further, columns three and four both highlight diversity in the ligand and metal cofactors associated with each protein. There is not a single protein described by having a single ligand; in fact, some have as many as five, and cofactors ranging from sulfate to Lutetium. Structural diversity among these 19 proteins also constitutes a unique feature of this problem.

Table 2 highlights some of the structural properties of the target proteins. Not only do the proteins differ in length, they also differ in CATH classification [16], which in turn describes variation in chain and domain characteristics for each protein. As an acronym, the Class, Architecture, Topology, and Homology expound on structural variation. Some proteins are primarily helical, others beta strand, and others mixtures of both; Table 3 provides cartoon rendering of these 19 proteins to further highlight their structural diversity.

Table 1. Target proteins described by organism, the ligands and metal complexes they bind. *1.Methanocaldococcus Jannaschii *2.Thermococcus Kodakarensis

PROTEIN	ORGANISM	LIGANDS	METALS
1A82	E. Coli mutant	ATP, DNN	MG
1ATP-E	House Mouse	ATP	MN
1E2Q	Human TMPK	ATP, TMP	MG
1F9A-C	M. Jannaschii*1	ATP	MG
1JJV	Human Flu	ATP, SO4	HG
1KP2-A	E. Coli	ATP, GAI, PO4	NONE
1MJH-A	M. Jannaschii*1	ATP	MN
1AYL	Plant E. Coli	ATP, OXL	MG
1B8A-A	T. K.*2	ATP	MN
1CSN	Fission Yeast	ATP, SO4	MG
1E8X-A	Wild Boar	ATP	LU
1G5T	Salmonella	ATP	MG
1GN8-A	E. Coli	ATP, SO4	MN
1HCK	Human	ATP	MG
1J7K	Thermotoga Maritima	ATP, ACT, HEZ	CO
1KAY	Cow	ATP	MG, CL, K
1NSF	Chinese Hamster	ATP	MG
1YAG	Human	ATP, SO4	MG
1PHK	European Rabbit	ATP	MN

Previously reported analyses [6] using Co and MolLoc performed a pairwise alignment of the 1ATP-E (PDB-id 1ATP, chain E) structure as the reference protein with each of the 18 remaining structures. Both methods ranked each of the 18 structures (not including 1ATP-E) based on the number of conserved atoms obtained for active site analysis. Table 4 shows the results of CO and MolLoc analysis ranked in the order of conserved atoms. Thus, proteins 1HCK, 1PHK, and 1CSN are ranked top three. These results are obtained with an average backbone RMSD of 1.76 and average SAS score of 10.04 [6]. Though orders vary, 1HCK, 1PHK, and 1CSN are the top three rankings for MolLoc as well. Results shown in Table 4 for MolLoc were obtained with an average backbone RMSD of 1.62 and average SAS score of 12.88 [6]. Structural Surface alignments aim to quantify results by obtaining the largest amount of corresponding atoms with the lowest relational RMSD and SAS scores. CO is seemingly better than MolLoc, even though in some cases they are closely comparable. These results are valuable since we too quantify results using backbone RMSD and our msTALI score [13] to quantify successful alignments.

2.2 Active site identification from msTALI

Active site identification of proteins from structural information can be an elusive pursuit due to its competing requirements. In one hand such a method should consider structural requisite to facilitate the enzymatic activity, while on the other hand be sensitive to the presence of a few (usually 3 or 4) critical residues in close proximity of each other without any attention on the conserved structure. Therefore a complete method operates both based on structure and sequence information. Our previous work has demonstrated the success of TALI [17] and msTALI [13] as such unique approaches. The msTALI is an extension of TALI by including multiple structure alignment in a manner that is analogous to ClustalW [18]. Our approach to structure alignment sets itself apart from other methods by including structural information such as: backbone torsion angles, atomic positions and membership of each residue in a secondary structural element, while including alignment of sequences using the Needleman-Wunsch [19] algorithm. The msTALI also includes other information such as water accessibility, structural information of side-chains (which are important in biochemistry of the enzyme), and properties of the neighboring atoms.

Here we utilize msTALI as a tool for active site identification through selection of structures with common enzymatic activity. We proceed based on the basic hypothesis that structure-sequence alignment of multiple proteins with common function will reveal the conserved regions (structural and sequence), which must contain the active site. Figure 1 provides an operational flowchart of our analysis procedure. In blue, our procedure describes the importance of classification and documentation. CATH [16] and PDB [20] are used for grouping proteins and verifying annotations for proteins that are documented. Thus, our method is validated by the common core. Further, the aim is that our method will contribute to documentation. Target proteins and then groups of our target sets are analyzed by msTALI as outlined in purple. Our returned results are analyzed and issued to our refinement process (colored blue and purple). The procedure then concludes with visualization, annotated phylogeny trees, and report generation (colored orange). Thus, the orange section describes the steps that verify the consistency of msTALI with respect to confirmed active sites.

Table 2. Table illustrating the target proteins, their sizes, and with their CATH classifications. C-class A-architecture T-topology H-homology

PROTEIN	Length (Res)	CATH CLASS
1A82	224	3.40.50.300
1ATP-E	350	1.10.510.10, 3.30.200.20
1E2Q	215	3.40.50.300
1F9A-C	168	3.40.50.620
1JJV	206	3.40.50.300
1KP2-A	455	3.40.50.620, 3.90.1260.10, 1.10.287.400
1MJH-A	162	3.40.50.620
1AYL	541	3.40.449.10, 2.170.8.10, 3.90.228.20
1B8A-A	438	2.40.50.140, 3.30.930.10
1CSN	298	3.30.200.20, 1.10.510.10
1E8X-A	961	3.10.20.90, 2.60.40.150, 1.25.40.70, 3.30.1010.10, 1.10.1070.11
1G5T	196	3.40.50.300
1GN8-A	159	3.40.50.620
1HCK	298	3.30.200.20, 1.10.510.10
1J7K	334	3.40.50.300, 1.10.10.10, 1.10.8.60
1KAY	381	3.30.420.40, 3.30.30.30, 3.30.420.40, 3.90.640.10
1NSF	273	3.40.50.300, 1.10.8.60
1YAG	375	3.30.420.40, 2.30.36.70, 3.30.420.40, 3.90.640.10, 3.40.20.10
1PHK	298	3.30.200.20, 1.10.510.10

Table 3. Cartoon rendering of the 19 target proteins.

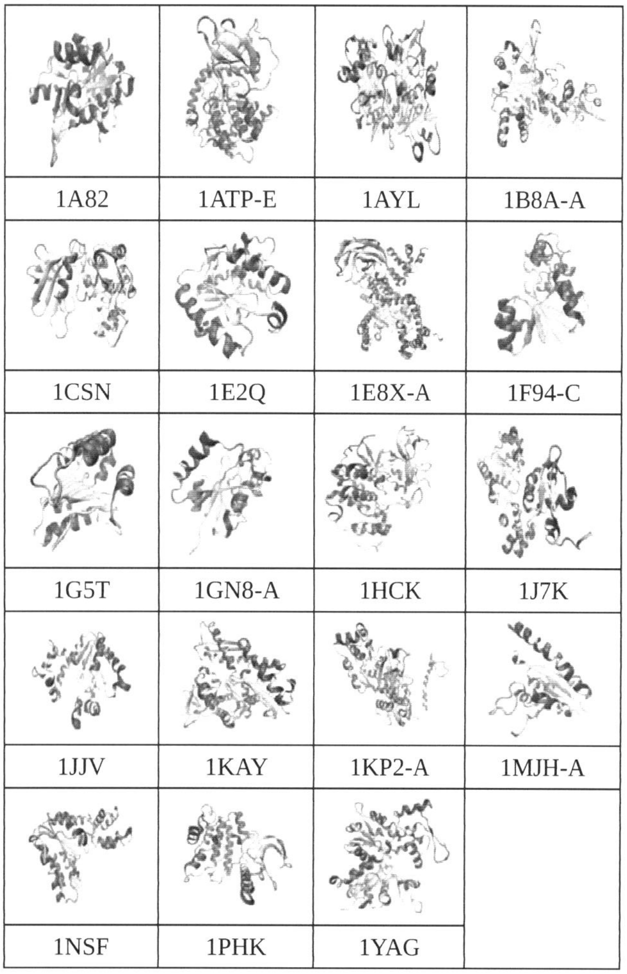

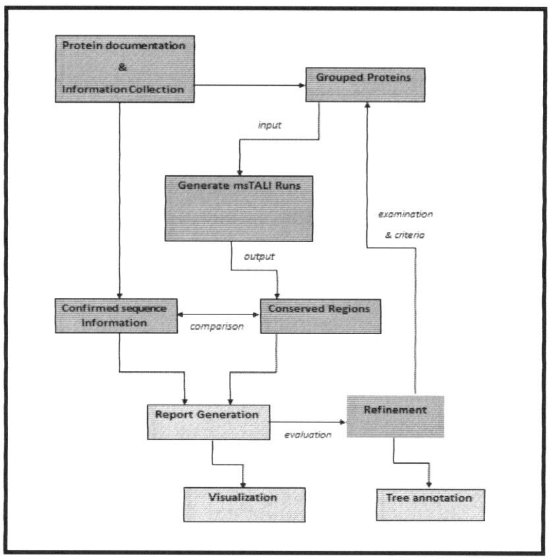

Figure 1. Methodical flow chart describing how msTALI runs are used to generate conserved regions, evaluate, refine, and validate results for reporting procedures.

Table 4. The number of conserved atoms identified by CO and MolLoc. Results are sorted based on CO outcomes.

Protein paired with 1ATP-E	N.Corresp. Atoms CO	N.Corresp. Atoms MolLoc
1HCK	62	45
1PHK	57	63
1CSN	50	55
1NSF	34	11
1J7K	25	25
1E8X-A	24	20
1F9A-C	21	18
1KAY	20	8
1YAG	20	17
1A82	19	13
1JJV	18	10
1GN8-A	17	14
1B8A-A	16	10
1MJH-A	16	14
1E2Q	15	5
1KP2-A	13	15
1AYL	12	16
1G5T	7	8

3 Results and discussion

Results of msTALI alignment are reported in this section. These results have been obtained by use of the web version of the msTALI that can be found on: http://ifestos.cse.sc.edu/mstali. The use of this service is free and only requires an initial user registration.

3.1 Pairwise alignment with msTALI

Although msTALI is capable of simultaneous alignment of multiple structures, our initial evaluation of msTALI focuses on pairwise alignment of proteins. This exercise is necessary to establish its performance compared to the prior exercises of pairwise structure alignment. As described in section *2.1*, previous investigations [6] focused on pairwise alignment of all 18 target proteins with respect to 1ATP-E. Results of a similar pairwise alignment using msTALI are shown in Table 5. The first column in this table is the PDB-ID of the protein when aligned with msTALI to the protein 1ATP-E. The second column of this table lists the number of residues that was found in common with 1ATP-E by msTALI. Note that our results are based on the number of amino acids where as in comparison to results reported by CO and MolLoc that report the number of atoms (and not residues). A very conservative scaling factor of 7 can be used to convert the number of residues to the number of atoms based on the number of common backbone atoms in every amino acid (N, H_N, C_α, H_α, C_β, C', O). Although msTALI discovers nearly an order of magnitude more number of atoms than the previously reported, the ranking of the most similar to the least similar structures (based on size of the conserved regions) remains very similar to that of CO and MolLoc. Protein structures 1CSN, 1HCK and 1PHK were consistently (by all three approaches) ranked as the top three homologous structures to 1ATP-E with 50, 62, and 57 conserved atoms reported by CO and 55, 45, and 63 conserved atoms reported by MolLoc. Additionally the average backbone RMSD for CO and MolLoc were 1.76 and 1.62. The average SAS [6] scores for CO and MolLoc were 10.04 and 12.88. While msTALI identifies the same three top ranking proteins as the most homologous structures, it discovers nearly an order of magnitude more number of conserved atoms (1,771, 1,701, and 1,785 atoms respectively) than previously reported. Figure 2 illustrates the successful pairwise alignments for the observed proteins. Moreover, describing a large amount of corresponding residues aligned with a low backbone RMSD. Next, we discuss analysis with respect to the structural variance of the targeted proteins by structural alignments on all 19 proteins simultaneously.

Table 5. Pairwise structural alignments using 1ATP-E with respect to the remaining 18 target proteins. Results are obtained using msTALI and listed in sorted order based on CO outcomes.

Protein paired with 1ATP-E	N.Corresp. Residues msTALI
1HCK	243
1PHK	255
1CSN	253
1NSF	51
1J7K	62
1E8X-A	19
1F9A-C	94
1KAY	26
1YAG	30
1A82	62
1JJV	46
1GN8-A	17
1B8A-A	71
1MJH-A	29
1E2Q	49
1KP2-A	56
1AYL	30
1G5T	36

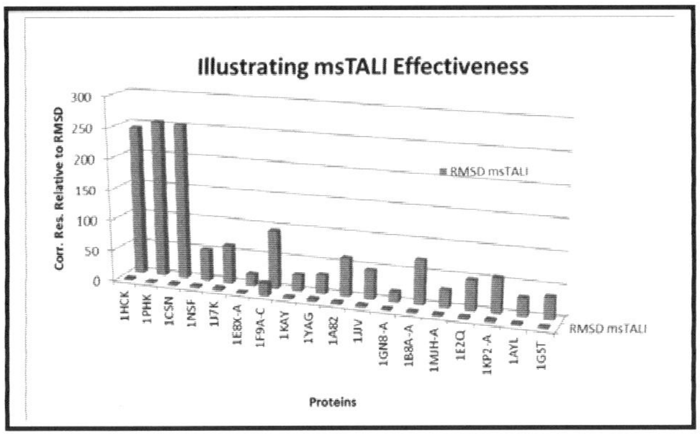

Figure 2. Performing pairwise comparisons across a dataset of proteins aligned with 1ATP-E using msTALI

3.2 Phylogenetic analysis of target proteins with msTALI

Before interpreting the results of multiple structure alignment with msTALI it is important to establish the legitimacy of its alignment. During the process of multiple structure alignment, msTALI produces the phylogeny of proteins, which can be interpreted as the clustering of structural similarities. Figure 3 illustrates the phylogenetic results of the msTALI analysis. As it can be seen, there is a clear agreement between the phylogenetic relationship established by msTALI and that of CATH classification. Although there is a clear correlation between the two sets of classification (CATH and msTALI), there are some observable differences. These differences occur for two main reasons. The first set of differences appears from the ambiguity in CATH's classification of some proteins. For instance the protein 1YAG can be classified in as many as five classes of proteins as listed in Table 2. The second set of differences is inherent to the questionable classification of proteins in some instances as previously reported [13].

The general similarity that is observed between msTALI's phylogeny results and CATH's classification of the 19 proteins substantiates the success of msTALI and allows reliable interpretation of its active site discovery.

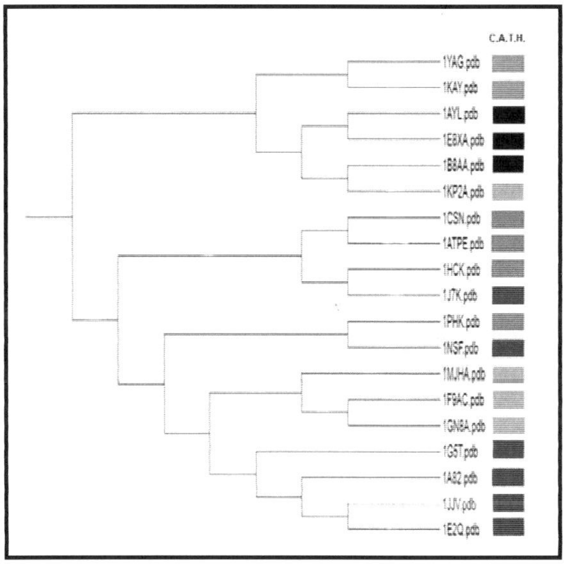

Figure 3. Phylogeny tree generated by using msTALI for a structural alignment on the 19 observed proteins simultaneously. Proteins marked with the same color are grouped based off of CATH classification

3.3 Active site identification with msTALI

It can be argued that an intersection of all the conserved residues listed in Table 5 could constitute the ATPase active site. However, such a simplistic approach may render an empty set as the intersection of the conserved residues. The establishment of the intersection region can be a difficult task since there may be little sequence similarity among the conserved regions. The similarity may be based primarily on structure and not sequence. Finally, it is possible that an entirely different set of conserved regions may surface if a structure other than the 1ATP-E was used as the reference structure during the pairwise alignment exercise. Multiple structure/sequence alignment feature of msTALI successfully addresses all the above concerns.

Table 6 shows the results of msTALI's multiple structure/sequence alignment. The second and third columns of this table list the conserved sequences across all the 19 structures, and the amino acid numbers respectively. The conversed regions identified by msTALI very closely correlate with the biologically confirmed active sites. Figure 4 provides an illustration of the msTALI identified active site and the biologically confirmed active site for the protein 1ATP-E. *Section A.* of the figure is the biologically confirmed active site. Further investigating the conserved region, the eight residues were described by exhibiting a coil structure connected to a helical region and all eight residues were partially accessible to the surface of the protein. Additionally, the core eight serve as a back-bridge for confirmed active site cavity openings and were essentially, structurally adjacent; this is picture in *section B.* of the figure. Last, *section C.* shows residues common in *A* and *B*, completing our description, while *section D.* shows the relationships of active sites across *A*, *B*, and *C* superimposed with one another.

Table 6. Result of multiple structure alignment using msTALI. Column two shows the conserved amino acid sequence for the residues reported in column three.

Protein	Conserved Residues	Conserved Residues
1A82	vgkt asca	13-16, 18-21
1ATP-E	ehtl ekri	86-89, 91-94
1E2Q	agks qsrk	17-20, 22-25
1F9A-C	pfhk hlev	11-14, 16-19
1JJV	sgkt ianl	13-16, 18-21
1KP2-A	ykgn ierf	131-134, 136-139
1MJH-A	fset eial	14-17, 19-22
1AYL	nyll lkgi	220-223, 225-228
1B8A-A	rpev aifk	130-133, 135-138
1CSN	pqlr eyrt	50-53, 55-58
1E8X-A	nlqi-ycql	430-435, 459-460
1G5T	kgkt aafg	39-42, 44-47
1GN8-A	pitn hidi	13-16, 18-21
1HCK	stai eisl	46-49, 51-54
1J7K	lvkq dmaa	87-90, 92-95
1KAY	davv smdk	80-83, 85-88
1NSF	wgdp trvl	510-513, 515-518
1YAG	pmnp snre	109-112, 114-117
1PHK	lrea lkev	66- 69, 71-74

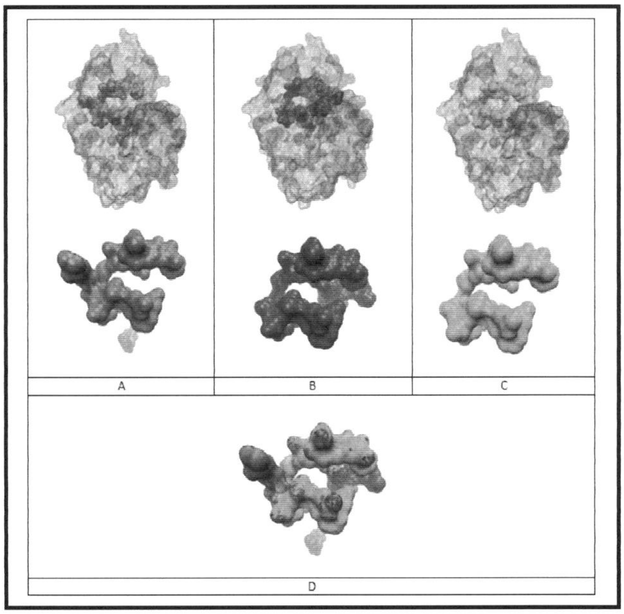

Figure 4. The active site for 1ATP-E. (A) Illustrates confirmed residues for ATP binding as documented from the PDB. (B) Illustrates conserved core residues responsible for ATP binding as output returned from msTALI. (C) Illustrates the overlapping areas commonly expressed from representations A and B. (D) Illustrates the active sites rendered in respect to one another.

4 Conclusion

Our application of msTALI for active site identification has successfully identified a motif characteristic to ATPase activity. Additionally, we reported the successful identification for the ATPase active site complex documented for 1ATP-E, while adhering to conventions of our phylogenetic analysis. Results convey a successful strategy in recovery of the active site in the case of ATPase activity. Though initial runs are successful, additional improvement to the strategy reported in this work can be imagined. The additional improvements can be based on clustered study of the proteins that is in accordance to the structural similarity shown in Figure 3. Further, similarity of charge and size of the cofactors can be considered in regrouping of the proteins.

Future investigations will also include extending the msTALI core engine. Studies utilizing other examples of enzymatic activity are to come and continued work addressing a point of convergence for the number of proteins aligned, as well as the number of required refinements will follow. Collectively, it is evident that the existing studies establishing our understanding of active sites stress the importance of sequence, structure, and biochemical properties of proteins in their function. The msTALI tool is unique compared to other existing software and meets these ideals.

5 Acknowledgments

Funding for this work is provided by ASPIRE-II grant from the University of South Carolina Research Foundation.

6 References

[1] M. Hendlich, F. Rippmann, and G. Barnickel, "LIGSITE: automatic and efficient detection of potential small molecule-binding sites in proteins," *J. Mol. Graph. Model.*, vol. 15, no. 6, pp. 359–363, Dec. 1997.

[2] D. G. Levitt and L. J. Banaszak, "POCKET: A computer graphies method for identifying and displaying protein cavities and their surrounding amino acids," *J. Mol. Graph.*, vol. 10, no. 4, pp. 229–234, Dec. 1992.

[3] B. Huang and M. Schroeder, "LIGSITEcsc: predicting ligand binding sites using the Connolly surface and degree of conservation.," *BMC Struct. Biol.*, vol. 6, p. 19, 2006.

[4] R. A. Laskowski, "SURFNET: A program for visualizing molecular surfaces, cavities, and intermolecular interactions," *J. Mol. Graph.*, vol. 13, no. 5, pp. 323–330, Oct. 1995.

[5] J. Liang, H. Edelsbrunner, and C. Woodward, "Anatomy of protein pockets and cavities: measurement of binding site geometry and implications for ligand design.," *Protein Sci.*, vol. 7, pp. 1884–1897, 1998.

[6] P. Bertolazzi, C. Guerra, and G. Liuzzi, "A global optimization algorithm for protein surface alignment.," *BMC Bioinformatics*, vol. 11, p. 488, Jan. 2010.

[7] A. Gutteridge, G. J. Bartlett, and J. M. Thornton, "Using A Neural Network and Spatial Clustering to Predict the Location of Active Sites in Enzymes," *J. Mol. Biol.*, vol. 330, no. 4, pp. 719–734, Jul. 2003.

[8] J. S. Fetrow, A. Godzik, and J. Skolnick, "Functional Analysis of the Escherichia coli Genome Using the Sequence-to-Structure-to-Function Paradigm : Identification of Proteins Exhibiting the Glutaredoxin / Thioredoxin Disulfide Oxidoreductase Activity," *J. Mol. Biol.*, pp. 703–711, 1998.

[9] J. Skolnick and J. S. Fetrow, "From genes to protein structure and function : novel applications of computational approaches in the genomic era," *TIBTECH*, vol. 18, no. January, pp. 34–39, 2000.

[10] A. Shulman-peleg, M. Shatsky, R. Nussinov, and H. J. Wolfson, "MultiBind and MAPPIS : webservers for multiple alignment of protein 3D-binding sites and their interactions," *Nucleic Acids Res.*, vol. 36, no. May, pp. 260–264, 2008.

[11] M. Jambon, O. Andrieu, C. Combet, G. Dele, C. Geourjon, and M. Sa, "Structural bioinformatics The SuMo server : 3D search for protein functional sites," vol. 21, no. 20, pp. 3929–3930, 2005.

[12] M. Punta, P. C. Coggill, R. Y. Eberhardt, J. Mistry, J. Tate, C. Boursnell, N. Pang, K. Forslund, G. Ceric, J. Clements, A. Heger, L. Holm, E. L. L. Sonnhammer, S. R. Eddy, A. Bateman, and R. D. Finn, "The Pfam protein families database.," *Nucleic Acids Res.*, vol. 40, no. Database issue, pp. D290–301, Jan. 2012.

[13] P. Shealy and H. Valafar, "Multiple structure alignment with msTALI.," *BMC Bioinformatics*, vol. 13, no. 1, p. 105, Jan. 2012.

[14] S. Angaran, M. E. Bock, C. Garutti, and C. Guerra, "MolLoc : a web tool for the local structural alignment of molecular surfaces," *Nucleic Acids Res.*, vol. 37, no. May, pp. 565–570, 2009.

[15] A. Kahraman, R. J. Morris, R. A. Laskowski, J. M. Thornton, and J. I. Centre, "Shape Variation in Protein Binding Pockets and their Ligands," *J. Mol. Biol.*, pp. 283–301, 2007.

[16] A. L. Cuff, I. Sillitoe, T. Lewis, O. C. Redfern, R. Garratt, J. Thornton, and C. A. Orengo, "The CATH classification revisited — architectures reviewed and new ways to characterize structural divergence in superfamilies," *Nucleic Acids Res.*, vol. 37, no. November 2008, pp. 310–314, 2009.

[17] X. Miao and M. G. Bryson, "TALI : Protein Structure Alignment Using Backbone Torsion Angles."

[18] J. D. Thompson, D. G. Higgins, and T. J. Gibson, "CLUSTAL W: improving the sensitivity of progressive multiple sequence alignment through sequence weighting, position-specific gap penalties and weight matrix choice.," *Nucleic Acids Res.*, vol. 22, no. 22, pp. 4673–4680, 1994.

[19] S. B. Needleman and C. D. Wunsch, "A general method applicable to the search for similiarities in the amino acid sequence of two proteins," *J. Mol. Biol.*, vol. 48, no. 3, pp. 443–453, 1970.

[20] M. D. Brice, J. R. Rodgers, and O. Kennard, "The Protein Data Bank," *Eur. J. Biochem*, vol. 324, pp. 319–324, 1977.

Protein Structure-Function Analysis with Self-Organizing Maps

Seonjoo Lim **Stephen Jaegle** **Lutz Hamel**
Department of Computer Science and Statistics
University of Rhode Island
Kingston, Rhode Island, USA
hamel@cs.uri.edu

Abstract

Here we describe an approach for protein structure-function analysis using self-organizing maps based on the structure of a protein's functional site. Our current approach differs from other approaches in that we directly unfold the 3D structure of the functional center of a protein into a suitable high-dimensional feature vector and then use self-organizing maps to discover similarities/dissimilarities between the corresponding feature vectors. We successfully applied our techniques to two large protein families: the protein kinases and the Ras superfamily. Even though a number of different approaches using self-organizing maps for the conformational analysis of molecules have been proposed, our approach is novel in that we apply it to protein families, use an efficient feature vector construction, and use a recently developed self-organizing map package that provides statistical support for evaluating the resulting map quality.

1 Introduction

The function of a protein is mainly determined by structural features, especially the functional site of the protein [1]. Therefore, common functionalities among proteins can be inferred from their structural similarities. Furthermore, in large protein families we can observe that small differences in the structure of the well preserved functional sites denote differences in functionality between the proteins. Here we describe an approach for protein structure-function analysis using self-organizing maps (SOMs) [2]. Our approach is based on the structure of a protein's functional site. The approach we employ here differs from other approaches, *e.g.* [3, 4, 5], in that we unfold the 3D structure of the functional center of a protein into a suitable high-dimensional feature vector and then use self-organizing maps to discover similarities/dissimilarities between the corresponding feature vectors. This approach to constructing feature vectors is substantially more efficient than the approach first outlined in [6]. Perhaps the work most closely related to ours is [7] where the authors classify protein motifs using SOMs. However, their work differs substantially from ours in that instead of directly encoding 3D spatial information of the motifs in question the authors compute a feature vector for a motif by looking at the angles at the α-carbon atoms along the backbone of a protein.

We successfully applied our technique to two large protein families: the protein kinases [8] and the Ras superfamily [9]. Our proposed approach seems to be novel in that we apply it to protein families, use an efficient feature vector construction, and use a recently developed self-organizing map package that provides statistical support for evaluating self-organizing map quality [10].

The remainder of this paper is structured as follows. In Section 2 we describe our methodology for aligning functional sites, extracting feature vectors, and computing SOM based models. We look at details of model building and cluster analysis in Section 3. In particular, we discuss the application of our technique to the two different protein families. Finally, we discuss our conclusions and further work in Section 4.

2 Preprocessing the Protein Structure Information

The major steps for preprocessing protein data are summarized in Figure 1. First, the protein structures for proteins under investigation are pulled from the Protein Data Bank (PDB) [11]. The proteins are then aligned using FATCAT [12]. From the aligned proteins we then extract only the functional site structures for our functional site based analysis. In order to filter out the functional sites, key structural information is used, like the consensus of a motif or the positional information (e.g., residue number) of a binding site for each protein. Next, the structures are simplified by extracting only the α-carbons from these functional sites. This provides information on the backbone structure of the functional sites by excluding the side chains. Finally, each functional site is represented by the 3D-coordinates of its α-carbons, and the coordinate data of all the α-carbons is unfolded into a linear vector – the feature vector of the functional site.

Figure 2 shows a set of feature vectors as rows. Each row

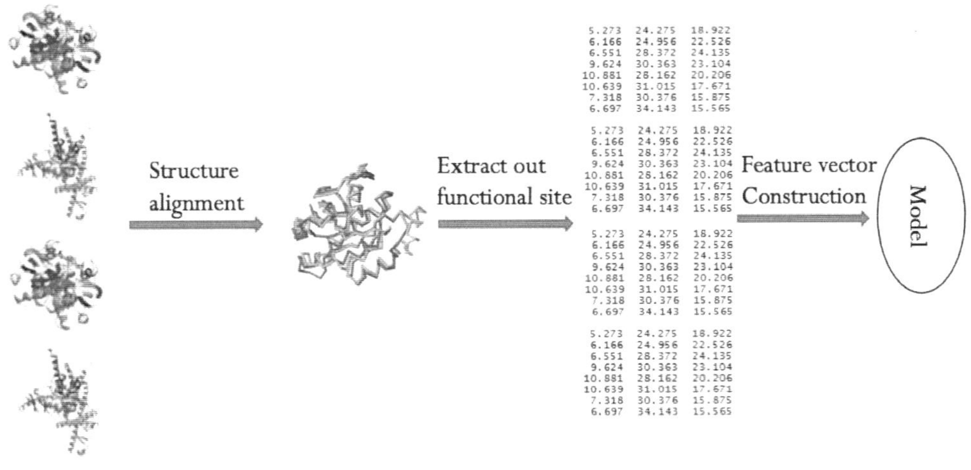

Figure 1: Steps in the protein structure-function analysis.

	family	ids	X1	X2	X3	X4	X5	X6	X7	X8	X9
1	HRas	121p	5.273	24.275	18.922	6.166	24.956	22.526	6.551	28.372	24.135
2	HRas	1qra	5.252	24.249	18.962	6.297	24.972	22.523	6.522	28.442	23.963
3	HRas	1ctq	5.278	24.256	18.976	6.262	24.946	22.534	6.503	28.426	23.990
4	HRas	1p2s	5.170	24.126	19.013	6.269	24.884	22.596	6.456	28.492	23.750
5	HRas	1agp	5.330	24.484	18.940	6.208	24.978	22.556	6.508	28.340	24.146
6	KRas	4dsn	5.348	24.380	18.886	6.154	24.970	22.557	6.604	28.328	24.302
7	RhoA	1a2b	5.149	24.282	18.878	6.373	24.932	22.453	6.699	28.276	24.223
8	RhoA	1cc0	5.153	24.122	18.987	6.432	24.981	22.549	6.441	28.517	23.924
9	RhoA	1cxz	5.202	24.334	18.906	6.462	24.972	22.434	6.659	28.312	24.221
10	RhoA	1dpf	5.378	24.265	18.894	6.376	25.021	22.497	6.459	28.517	24.011
11	RhoA	1ftn	5.153	24.122	18.988	6.432	24.981	22.548	6.441	28.517	23.925
12	Rab1A	2fol	5.217	24.048	18.990	6.134	24.992	22.576	6.489	28.664	23.592
13	Rab1A	2wwx	5.564	24.206	18.775	6.150	25.179	22.363	6.372	28.705	23.726
14	Rab1A	3sfv	5.264	24.036	19.057	6.325	25.017	22.610	6.309	28.671	23.678
15	Rab1A	3tkl	5.165	24.113	19.010	6.364	25.004	22.511	6.405	28.514	23.895
16	Rab1B	3jza	5.434	24.280	18.825	6.028	25.179	22.421	6.506	28.721	23.680
			G(x,y,z)			X(x,y,z)			X(x,y,z)		

Figure 2: Feature vector construction, unfolded 3D-coordinates.

represents the residues of a functional site of a protein from the Ras family. For demonstration purposes the functional sites have been truncated to three amino acids. In our actual setting we consider eight residues. Here each feature vector is denoted by two labels (a family name and a PDB ID), and three sets of attributes representing the 3D coordinates of the α-carbon of the three residues (GXX). The three residues are the first three of the eight residues making up the phosphate binding loop (p-loop) motif GXXXXGK[S/T]. The p-loop is the active site in the Ras family. In other words, a feature vector for a protein is the sequential listing of the 3D coordinates of the α-carbons appearing in its functional site.

3 Model Building and Evaluation

For our experiments we used proteins from two large protein families: the Ras family and the protein kinase family. We preprocessed the proteins as described in Section 2 and then constructed self-organizing maps for each protein family. The maps shown are fully converged, i.e., the clusters, their size, and their relative position to each other are statistically meaningful [13]. We commence this section by briefly reviewing self-organizing maps.

3.1 Self-Organizing Maps

A self-organizing map [2] is a kind of artificial neural network that implements competitive learning, which can be considered a form of unsupervised learning. On the map itself, neurons are arranged along a rectangular grid with dimensions x_{dim} and y_{dim}. Learning proceeds in two steps for each training instance $\vec{x}_k$, $k = 1, 2, 3, \ldots, M$, with M the number of training instances:

1. The **competitive step** where the best matching neuron for a particular training instance is found on the rectangular grid,

$$c = \underset{i}{\operatorname{argmin}}(||\vec{m}_i - \vec{x}_k||)$$

where $i = 1, 2, \ldots, N$ is an index over the neurons of the map with $N = x_{dim} \times y_{dim}$ the number of neurons on the grid, and $\vec{m}_i$ is a neuron indexed by i. Finally, c is the index of the best matching neuron $\vec{m}_c$ on the map.

2. The **update step** where the training instance $\vec{x}_k$ influences the best matching neuron $\vec{m}_c$ and its neighborhood. The update step can be represented by the following update rule for the neurons on the map,

$$\vec{m}_i \leftarrow \vec{m}_i - \eta \vec{\delta}_i h(c, i)$$

for $i = 1, 2, \ldots, N$. Here $\vec{\delta}_i = \vec{m}_i - \vec{x}_k$, η is the learning rate, and $h(c, i)$ is a loss function with,

$$h(c, i) = \begin{cases} 1 & \text{if } i \in \Gamma(c), \\ 0 & \text{otherwise,} \end{cases}$$

where $\Gamma(c)$ is the neighborhood of the best matching neuron $\vec{m}_c$ with $c \in \Gamma(c)$. Typically the neighborhood is a function of time and its size decays during training. Initially the neighborhood for neuron $\vec{m}_c$ includes all other neurons on the map,

$$\Gamma(c)|_{t=0} = \{1, 2, \ldots, N\}.$$

As training proceeds the neighborhood for $\vec{m}_c$ shrinks down to just the neuron itself,

$$\Gamma(c)|_{t \gg 0} = \{c\}.$$

Here, as before, $N = x_{dim} \times y_{dim}$ is the number of neurons on the map. This means that initially the update rule for each best matching neuron has a very large field of influence which gradually shrinks to the point that the field of influence just includes the best matching neuron itself.

The two training steps above are repeated for each training instance until the given map converges.

Here we use our `popsom` package [10] which supports statistical convergence criteria [13] and detailed cluster visualizations in terms of our starburst plots [14]. Figure 3 shows a scatter plot of the hepta problem in Ultsch's fundamental clustering problem suite [15]. The data set consists of seven distinct clusters embedded in three dimensional space. Notice that there is a single, very tight cluster at (0,0) and then we have six clusters surrounding this center cluster. Figure 4 shows a SOM starburst plot of this data set. The seven clusters can easily be identified on the map by their starbursts. Also easily visible is the fact that clusters themselves are identified by their light color and cluster boundaries are identified by darker colors. The easily identified borders mean that the clusters are indeed distinct clusters. Their relative position is also meaningful to a point, given that this is a 2D rendering of a higher dimensional space. Here we see the cluster with class label 1 towards the center of the map. This is a representation of the tight center cluster in the original plot. We can also see that it consumes somewhat less map real estate than the other clusters meaning that the cluster is very tight. All these observations are justified due to the fact that the map has converged and therefore positioning and distance amongst clusters is statistically meaningful.

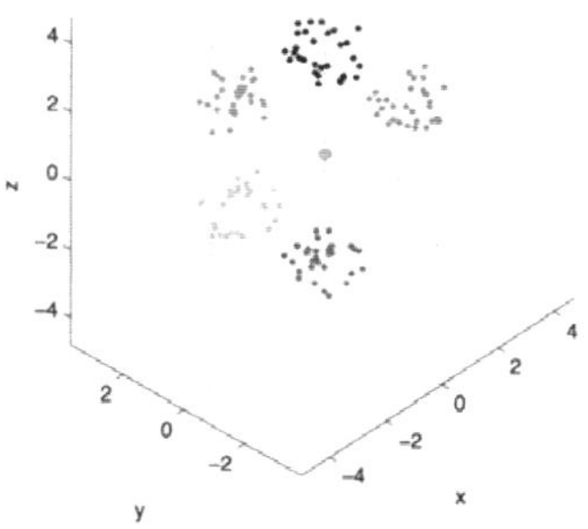

Figure 3: The FCPS Hepta data set.

Table 1: Hierarchy of the STE Kinase Family and corresponding Binding Sites.

Family	Subfamily	PDB ID	Binding Site
STE 7	MAP2K4	3ALO	108-116
STE 11	MAP3K5	4BF2 3VW6	686-694
STE 20	PAK6 PAK4	4KS7 2J0I, 4JDI	413-421

3.2 The Protein Kinase Family

Protein kinases catalyze proteins by attaching phosphate groups to them. For example, protein kinase helps bind ATP to proteins so that they can be phosphorylated and produce ADP. Here we consider the sterile (STE) group which is one of ten human kinase families [8]. Three main families in the STE group operate on each other sequentially: STE 20 activates STE11 and STE11 activates STE 7. Table 1 shows the STE families with their subfamilies and their corresponding members. Also shown is the binding site for each member. Note that the length of the respective binding sites is eight residues. As mentioned above, that means the corresponding feature vectors have a length of twenty four.

Figure 5 shows that the SOM algorithm was able to recover the subfamilies given in Table 1 as separate clusters. Here we make use of the fact that cluster relative positioning and the coloring of the map is statistically meaningful because the map has converged. We can observe that the clusters for

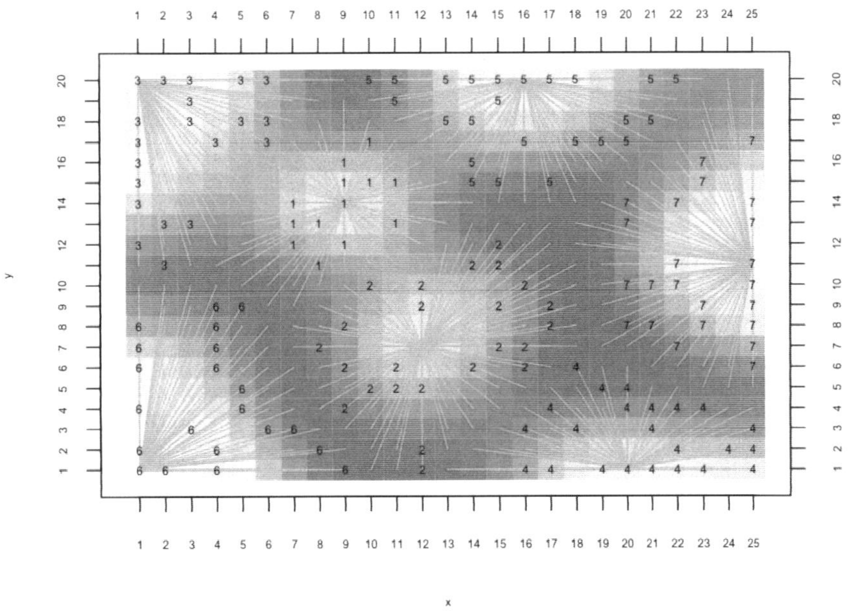

Figure 4: A SOM starburst plot of the Hepta data set.

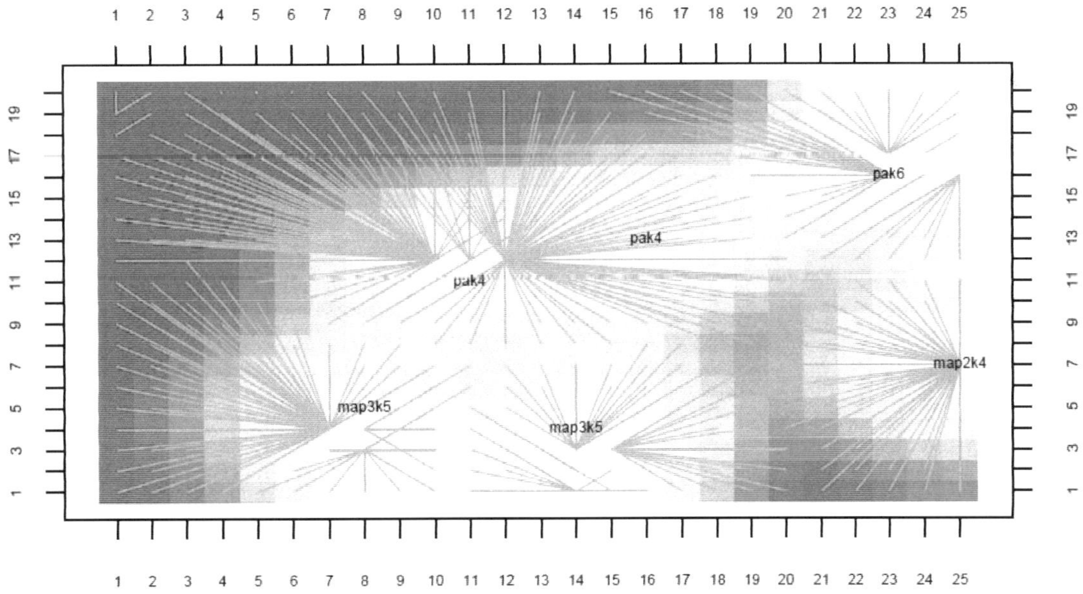

Figure 5: A SOM depicting the protein kinase STE Group.

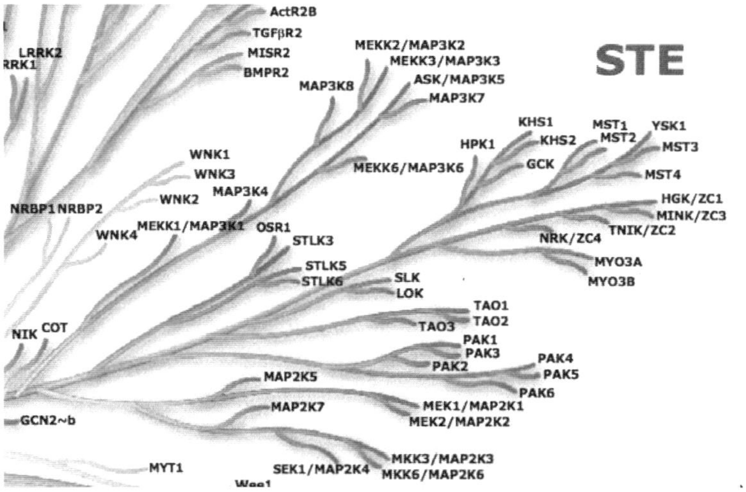

Figure 6: The STE branch of the evolutionary tree of the human kinase complement [8].

PAK4 and MAP3K5 are distinct but close together. We can therefore infer that the active site structures for those proteins are very similar. The same holds for the clusters PAK4 and PAK6. It is perhaps remarkable that the active site structure for MAP2K4 seems to share more similarity with PAK6 than with MAP3K5. It is tempting to see if an analysis solely based on the functional site of these proteins can recover the evolutionary relationships between the protein families shown in Figure 6 which is the STE branch of the evolutionary tree of the human kinase complement published as a supplement to [8]. Now, if we consider Figure 6 we can find MAP2K4, PAK4, and PAK6 on the lower branches of the tree. We can find MAP3K5 on an upper branch near the STE label. Tracing the evolutionary lines in Figure 6 it is clear that SOM's cluster structure captures the the evolutionary relationships with the exception of perhaps the MAP3K5/PAK4 relationship which on the SOM appears to be much closer than the tree indicates. One interpretation is that the structure of the active site of these two protein families has been highly preserved and that evolutionary differences manifest themselves in different protein domains.

3.3 The Ras Superfamily

The Ras superfamily of small GTPases is a large and diverse group of proteins that act as molecular switches for regulating cellular functions [9]. This superfamily is divided into five major families based on their structural and functional similarities: Rho, Ras, Rab, Ran, and Arf [16]. The protein members of the Ras superfamily have 40% - 85% of high primary sequence identity, while each subfamily has individual functions and different targets [17]. All members of the Ras superfamily have highly conserved common structural cores and function as GDP/GTP-regulated molecular switches. For example, a GTP-binding protein binds to either guanosine diphosphate (GDP) or guanosine triphosphate (GTP) so the protein becomes either inactive or active, respectively [18].

Table 2: Hierarchy of the Ras superfamily and the list of proteins used in the analysis.

Family	Subfamily	PDB ID
Ras	HRas	121P, 1QRA, 1CTQ, 1P2S, 1AGP
	KRas	4DSN
Rho	RhoA	1A2B, 1CC0, 1CXZ, 1DPF, 1FTN
Rab	Rab1A	2FOL, 2WWX, 3SFV, 3TKL
	Rab1B	3JZA
Arf	Arf1	1HUR
	Arf2	1U81
	Arf3	1RE0
	Arf4	1Z6X
Ran		1I2M, 1IBR, 1RRP, 3CH5, 3EA5, 3GJ3

There is a particular motif in the proteins of the Ras superfamily that determines the features of each subfamily. Each subfamily acts as a molecular switch for a unique target or intervenes in a cell process, such as cell proliferation. Members of this superfamily conserve five G domains which are fundamental subunits: G1-G5 [9]. G domains are highly conserved regions related to nucleotide binding, a process that is involved with the GDP/GTP cycle. The G1 domain contains the phosphate binding loop (p-loop), which is a common motif in GTP binding proteins with a consensus of GXXXXGK[S/T], where X denotes any amino acid and S/T means S or T. Interestingly enough, the length of the active site of the protein family measure also eight residues. Table 2 shows the hierarchical relationship of the Ras superfamily and the list of PDB IDs chosen for analysis in this paper.

Figure 7 shows a SOM constructed for the Ras superfamily.

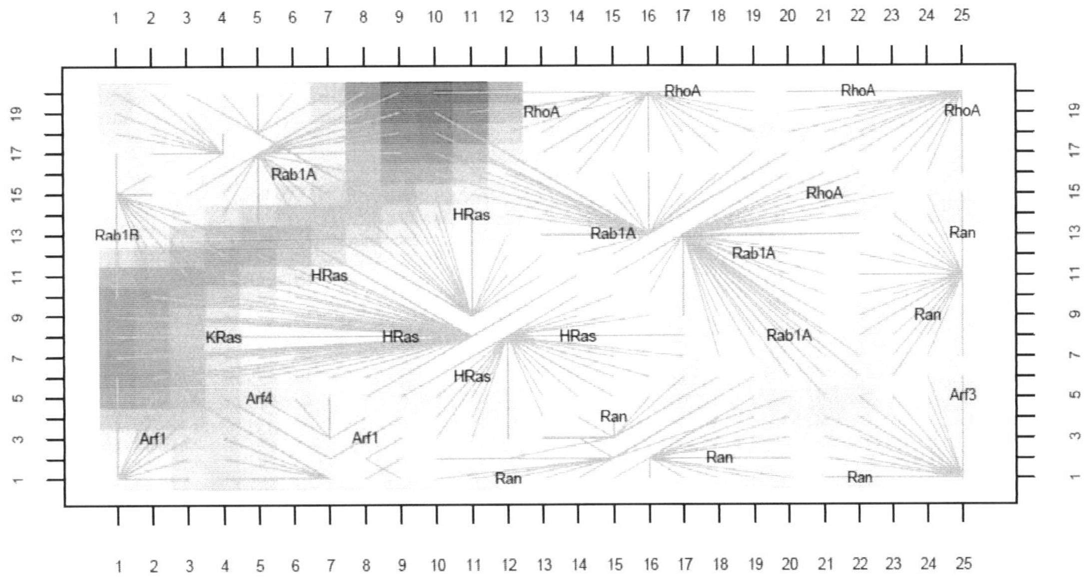

Figure 7: A SOM depicting the Ras Superfamily.

The major clusters for Rho (top-right), Rab and Ran (center-right), and Arf and Ras (bottom-center-left), are easily identified. What is curious is that there is a separate Rab cluster on the top-left of the map separated from the remaining clusters by a fairly dark border. This means that structurally these Rab proteins look substantially different from the remaining proteins. We intend to follow up and investigate.

We can now investigate whether an structure-function analysis solely based on the active site of the proteins preserves the evolutionary relationship of the proteins. Figure 8 shows a consensus tree of the Ras superfamily published in [19]. From the tree it is easily identified that the families Rab, Ran, and Rho are closely related to each other and that the families Arf and Ras form another cluster. Going back to our map in Figure 7 we can see that the relative positioning of the clusters on the map preserves these evolutionary relationships. We can observe one outlier - a sole Arf protein shows up in the Ran cluster at the bottom right of the corner.

4 Conclusions and Further Work

Here we described our approach for protein structure-function analysis using self-organizing maps based on the structure of a protein's functional site. Our current approach differs from other approaches in that we directly encode the 3D structure of the functional center of a protein into a suitable high-dimensional feature vector and then use self-organizing maps to discover similarities/dissimilarities between the corresponding feature vectors. We used our `popsom` package which supports statistical convergence and quality measures and advanced visualization techniques for self-organizing map construction and evaluation. We successfully applied our techniques to two large protein families: the protein kinases and the Ras superfamily. We have shown that SOM preserves protein intra-family relationships as clusters and inter-family relationships with the relative positioning of family cluster to each other on a map. We have shown that evolutionary relationships between protein families are to a large degree preserved within the active site of the proteins.

Future research will focus on applying this technique to other protein families and structures. We also intend to investigate the outliers we found on the map for the Ras super family.

References

[1] H. A. Maghawry, M. G. Mostafa, M. H. Abdul-Aziz, and T. F. Gharib, "Structural protein function prediction-a comprehensive review.," *International Journal of Modern Education & Computer Science*, vol. 7, no. 10, 2015.

[2] T. Kohonen, *Self-organizing maps*. Springer Berlin, 2001.

[3] M. T. Hyvönen, Y. Hiltunen, W. El-Deredy, T. Ojala, J. Vaara, P. T. Kovanen, and M. Ala-Korpela, "Application of self-organizing maps in conformational analysis of lipids," *Journal of the American Chemical Society*, vol. 123, no. 5, pp. 810–816, 2001.

[4] T. Murtola, M. Kupiainen, E. Falck, and I. Vattulainen, "Conformational analysis of lipid molecules by self-organizing

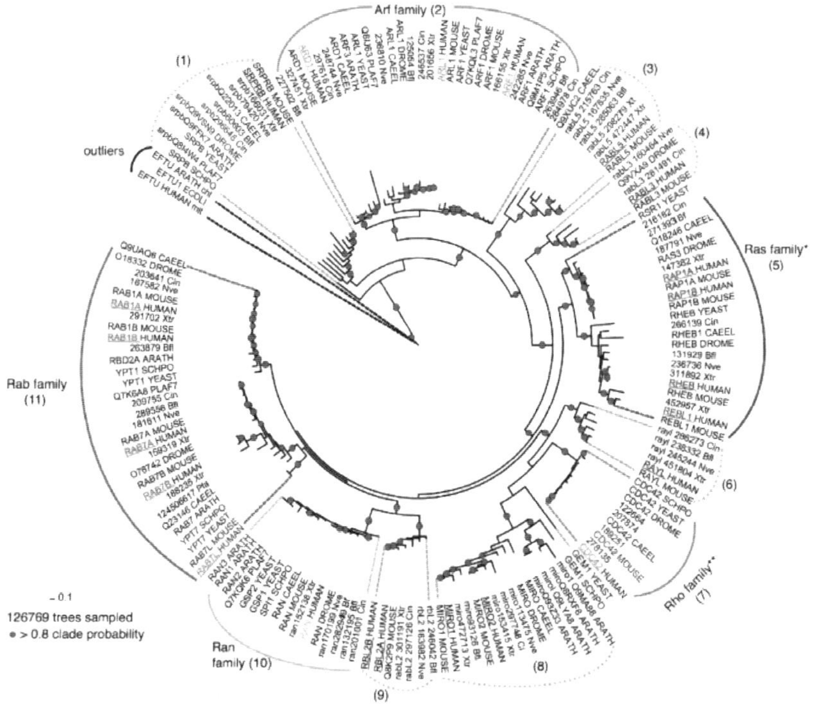

Figure 8: A consensus tree of the Ras Superfamily.

maps," *The Journal of chemical physics*, vol. 126, no. 5, p. 054707, 2007.

[5] D. Fraccalvieri, A. Pandini, F. Stella, and L. Bonati, "Conformational and functional analysis of molecular dynamics trajectories by self-organising maps," *BMC bioinformatics*, vol. 12, no. 1, p. 1, 2011.

[6] L. Hamel, G. Sun, and J. Zhang, "Toward protein structure analysis with self-organizing maps," in *Computational Intelligence in Bioinformatics and Computational Biology, 2005. CIBCB'05. Proceedings of the 2005 IEEE Symposium on*, pp. 1–8, IEEE, 2005.

[7] J. Schuchhardt, G. Schneider, J. Reichelt, D. Schomburg, and P. Wrede, "Local structural motifs of protein backbones are classified by self-organizing neural networks," *Protein engineering*, vol. 9, no. 10, pp. 833–842, 1996.

[8] G. Manning, D. B. Whyte, R. Martinez, T. Hunter, and S. Sudarsanam, "The protein kinase complement of the human genome," *Science*, vol. 298, no. 5600, pp. 1912–1934, 2002.

[9] K. Wennerberg, K. L. Rossman, and C. J. Der, "The ras superfamily at a glance," *Journal of cell science*, vol. 118, no. 5, pp. 843–846, 2005.

[10] L. Hamel, B. Ott, and G. Breard, *popsom: Self-Organizing Maps With Population Based Convergence Criterion*, 2015. R package version 3.0.

[11] P. W. R. et al., "The RCSB Protein Data Bank: new resources for research and education," *Nucleic acids research*, vol. 41, no. D1, pp. D475–D482, 2013.

[12] Y. Ye and A. Godzik, "Flexible structure alignment by chaining aligned fragment pairs allowing twists," *Bioinformatics*, vol. 19, no. suppl 2, pp. ii246–ii255, 2003.

[13] L. Hamel, "Som quality measures: An efficient statistical approach," in *Advances in Self-Organizing Maps and Learning Vector Quantization*, pp. 49–59, Springer, 2016.

[14] L. Hamel and C. W. Brown, "Improved interpretability of the unified distance matrix with connected components," in *7th International Conference on Data Mining (DMIN'11)*, pp. 338–343, 2011.

[15] A. Ultsch, "Clustering wih som: U* c," in *Proceedings of the 5th Workshop on Self-Organizing Maps*, vol. 2, pp. 75–82, 2005.

[16] R. Kahn, C. Der, and G. Bokoch, "The ras superfamily of gtp-binding proteins: guidelines on nomenclature.," *The FASEB journal*, vol. 6, no. 8, pp. 2512–2513, 1992.

[17] M. Kennedy, H. Beale, H. Carlisle, and L. Washburn, "Achieving signalling specificity: the ras superfamily.," *Nature Reviews Neuroscience*, vol. 6, pp. 423 – 434, 2005.

[18] B. Alberts, A. Johnson, J. Lewis, M. Raff, K. Roberts, and P. Walter, *Molecular Biology of the Cell.* Garland Science, 2002.

[19] A. M. Rojas, G. Fuentes, A. Rausell, and A. Valencia, "The ras protein superfamily: evolutionary tree and role of conserved amino acids," *The Journal of cell biology*, vol. 196, no. 2, pp. 189–201, 2012.

A Support Vector Machine Based Model for Predicting Heparin-Binding Proteins Using XB Patterns as Features

Joseph Sirrianni, Zhichun Xiao, Wing Ning Li
Computer Science and Computer Engineering, University of Arkansas, Fayetteville, Arkansas, U.S.A

Abstract - *Heparin is a highly sulphated and negatively charged polysaccharides belonging to the glycosamino-glycans(GAGs) family, used in medical treatments as an anticoagulant. Although many heparin-binding proteins have been identified, there are still many proteins needing to be classified as heparin-binding or not. Many studies have been aimed at prediction of heparin binding patterns in the primary structure of proteins, however, still no model has emerged which reasonably predicts proteins as heparin-binding or not. The main objective of this study is to predict heparin-binding proteins from their amino acid sequence information. A supervised learning algorithm based on support vector machine (SVM) is applied to two data sets; one training set used to create the model and one testing set used to validate and test accuracy of the model. For the testing set, the model achieves 75.36% accuracy in predicting heparin-binding proteins. The current model uses 66 XB patterns as features.*

Keywords: Heparin, Heparin-binding proteins, Heparin-binding Motifs, Support vector machine, Prediction model.

1 Introduction

Heparin is a member of the glycosaminoglycan (GAG) family. Glycosaminoglycans are linear polysaccharides that participate in many biological processes by interacting with a wide range of proteins [1]. Heparin is a highly sulfated glycosaminoglycan found in most organisms and is widely used as an anticoagulant to treat thrombosis, thrombophlebitis, and embolism [2].

There is a strong interest in finding new heparin- binding proteins and peptide sequences in order to develop new treatment methods for many diseases. To support the above task, we have applied the Support Vector Machine (SVM) [3-5] approach, a supervised machine learning method, to build a prediction model. The model takes in the primary structure (amino acid sequence information) of a protein and determines if the protein is heparin-binding or not.

2 Background

2.1 Biological background

Proteins are comprised of amino acids. These amino acids are strung together in an amino acid chain, which makes up the primary structure of a protein. The primary structure contains a lot of relevant data pertaining to the features and characteristics of its protein. An amino acid chain is a specific consecutive sequence of amino acids found in a protein. There are only 20 natural distinct amino acids found in proteins. From a computer science standpoint, a protein's primary structure is often represented in the format of a string of capital letters, where each letter maps to one of the 20 natural amino acids. For example, the amino acid sequence Alanine-Cysteine-Alanine-Glycine would correspond to the following string 'ACAG', where Alanine maps to the letter 'A', Cysteine maps to the letter 'C', and Glycine maps to the letter 'G'.

Available structural information on heparin-binding proteins (HBPs) reveals that heparin binds to a binding pocket consisting of positively charged amino acids (lysine/arginine/histidine) [6]. An XB pattern is a string of X and B, where B stands for the 3 basic amino acids (lysine/arginine/histidine) and X stands for the remaining 17 of the 20 natural amino acids. Based on information given by the structures of fibroblast growth factors (FGFs) (proteins that interact with heparin during their cell signaling process), it is known that the selective distribution of the basic amino acids is important for heparin affinity and interaction. Based on the 3-dimensional structures of other heparin-binding proteins, consensus or signature heparin-binding sequences (strings) have been found to occur in these proteins that are thought to be required for their interaction with heparin. The two main pattern strings are XBBXBX and XBBBXXBX [7]. These two patterns and another 17 patterns are further investigated in their occurrences in heparin-binding proteins [8], as are other commonly occurring patterns.

2.2 Computer science background

In the SVM based Supervised Machine Learning, samples are considered as points in a higher dimensional space. In this case, the samples will be individual proteins. Each point has a label that indicates to which of the two groups it belongs (in this case heparin-binding protein group and non-heparin- binding protein group). Further, a set of training or learning samples (usually half of the samples belong to one group and another half belongs to the other group) are fed into the SVM learning algorithm. The algorithm attempts to build a hyper plane that separates the learning samples into the two groups. Samples belonging to the same group can be expected to reside on one side of the hyper plane and consequently the hyper plane becomes the classifier or the prediction model. It should be noted that any hyper plane partitions a higher dimensional space into two parts, on either sides of the hyper plane. To predict a new sample, the sample is transformed into a point in the higher dimensional space. Then depending on which side of the hyper plane the point ends up on, a prediction is made.

In a higher dimensional space, each dimension is a feature of the underlining sample (in this case proteins are samples). In building a SVM based prediction model for heparin-binding proteins, we need to decide what features of proteins to use. In this study, we have decided to use occurrences of different XB patterns as features, since, as stated above, the XB pattern occurrences in heparin- binding proteins seem to suggest that their presence is important to the potential to bind.

3 Contribution

Using XB patterns as features, a SVM based prediction model for heparin-binding proteins is proposed and developed. The prediction is based on sequence information or protein primary structure. The models achieve reasonable prediction results and support the research effort of finding new heparin- binding proteins and peptide sequences in order to develop new treatment methods for many diseases. As of now, we are unaware of any other heparin-binding predictive models existing currently.

4 Approach

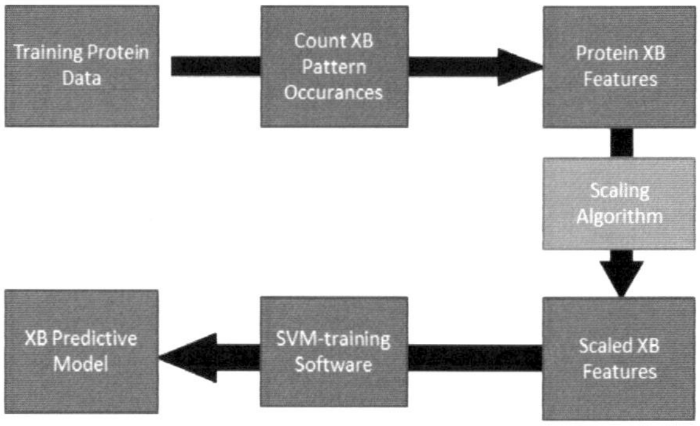

Figure 1. The process of creating the predictive model from the initial training proteins.

The total process of creating the model is displayed in figure 1. The initial amino acid chain in FASTA format is converted into feature values. Then the new feature values are scaled and used to create the predictive SVM model.

4.1 Data sample collection

The protein's primary structure information used for training and testing were extracted from the UniProt database [13]. One hundred and seventy four heparin binding proteins were gathered from the database of polyanion-binding proteins (DB-PABP) [16], and another one hundred and four non-binding proteins were gathered from the UniProt database. Tables 1 and 2 list examples of heparin binding and non-heparin binding proteins used in the testing and training. A complete list of the proteins used can be found in the appendix.

Table 1. Examples of heparin binding proteins used in testing and training

Sequence Identifier	Description
731717	Protein NSG1
307129	hepatic lipase precursor
33959	interleukin 8 [Homo sapiens]
1585498	diamine oxidase
340146	urokinase

Table 2. Examples of non-heparin binding proteins used in testing and training

Sequence Identifier	Description
O35400	alcohol sulfotransferase
Q15125	cholestenol DELTA-isomerase
Q08462	adenylate cyclase
P08842	steryl-sulfatase

P25697	phosphoribulokinase

The training set for the models consisted of 140 total proteins, which is about half of the total protein sample size. The remaining proteins were part of the testing set, which was used to determine accuracy. Of the 140 training proteins, 70 were known to be heparin binding and the other 70 were known to be not heparin binding. The training set was evenly distributed between binding and non-binding proteins in order to ensure the models would be balanced in their construction.

4.2 Feature selection

A total of 66 features were used to map each protein to a point in a 66 dimensional space. The feature values are calculated based on the number of occurrences of each feature's corresponding XB pattern in the protein's primary amino acid sequence. Therefore, our model only uses a protein's primary sequence data to make the heparin binding prediction.

Each protein has a primary structure, which is its sequence of amino acids. Each amino acid has its own generalized pH level. Heparin is one of the mostly negatively charged glycosaminoglycans, so it's hypothesized that a larger volume of basic amino acids are necessary in a protein for it to bind. A protein's sequence can be reduced to the two symbols B and X, where B represents the 3 basic amino acids (arginine, histidine, and lysine) and X represents the 17 non-basic amino acids [2, 9, 10, 11]. Recent studies have shown that heparin-binding proteins contain common XB pattern strings [8]. Thus, the XB patterns present in a protein should be relevant to its potential to bind. The XB patterns selected for the model are patterns that have been identified as being present in other heparin binding proteins.

When deriving the XB features of a protein, it first has its primary structure converted from a string of amino acid identifiers to a string of just Xs and Bs. Then, each of the 66 XB pattern occurrences is individually summed and set to the corresponding feature number (Feature 1 is the first XB pattern, Feature 2 is the second, and so on). A list of the first 7 features and their corresponding XB patterns is listed in Table 3. A full list of the 66 XB patterns can be requested from the authors.

Table 3. Example set of Features 1 – 7 and their XB patterns

Feature Number	XB pattern
1	XBXXBXBX
2	XBXBXXBX
3	XBBXBBX
4	XBBBXXBX
5	BBXXBBXX
6	XBBXXBBX
7	BXBXBBXB

The software used for feature derivation consisted of python scripts. The python scripts transformed the protein data from FASTA format into the desired feature values.

4.3 Scaling the features

Every feature is represented by a number. In this case, the XB features count occurrences of XB patterns, so their feature value must be a positive integer. However, some XB patterns are much smaller than others and occur more often; this can result in some features values being much greater than others and overshadowing them in importance.

To handle the various possible ranges of the feature values, all of the features were scaled to fall within a range of [-1, +1]. This scaling helps the support vector machine balance out the importance of each feature, to avoid features with large values dominating ones with smaller values [3].

4.4 The support vector machine

In addition to the selected features, several aspects of the support vector machine affect the accuracy of the model. These aspects include the kernel function type, the function parameter, γ, and the soft margin parameter, C.

The kernel function translates the training data into higher dimensions so that the training points can become linearly separable. After testing several different kernel functions, the kernel function chosen for the SVM was the Gaussian radial basis function. This kernel takes in a specified parameter γ. The Gaussian function is as such:

$$K(x_i, x_j) = \exp\left(-\gamma * |x_i - x_j|^2\right), \quad \gamma > 0. \quad (1)$$

The soft margin parameter, C, (also called the penalty parameter) determines how much variability in the computed hyperplane should be allowed. A larger C value will result in a tighter bound, while a smaller C value would result in a more relaxed bound [4].

In order to help determine optimal γ and C values, the grid tool from LIBSVM, a support vector machine library, was used [15]. The grid tool performs programmatic cross-validation checks with various combinations of γ and C values to determine which yields the best accuracy. These cross-validation checks are performed by separating the training data into several subsets and then running a test in which half of the subsets are considered the training set and the other half is considered the testing set. The cross validation model is created using the training subsets and tested against the testing subsets in order to achieve accuracy [12]. The values of (γ, C) which yielded the highest accuracy were used in the models.

4.5 Model validation

In order to validate the model a testing protein data set was identified consisting of the remaining 138 proteins not included in the training set. Of these, 105 of the proteins were known heparin binding proteins, while the other 34 were known heparin non-binding proteins.

A different set of LIBSVM functions was used to validate the model. The validation was performed by first transforming the testing protein's sequence data into the features used by the model, scaling them, and then determining on which side of the dividing hyperplane each protein is located. The proteins were then assigned a label, either +1 or -1, to signify which set the protein was predicted to fall into, where the +1 set is the set of heparin binding proteins and the -1 set is the set of heparin non-binding proteins. Each newly predicted label is then compared to the protein's known label. If the labels match, then the prediction is considered successful. If the labels do not match, the prediction is considered unsuccessful.

5 Results

For this research a model using 66 XB pattern features was created. The results of its accuracy are listed in Table 4 below.

Table 4. Model Accuracies and parameter values.

66 XB Pattern Features Model	
γ value	8
C value	0.03125
Training Set Accuracy	99.28% (139/140)
Testing Set Accuracy	75.36% (104/138)
Combined Accuracy	87.41% (243/278)

The model tended to do well (close to 100% accuracy) predicting the Training set. The Testing set was predicted with a high accuracy of 75.36%. The combined accuracy (combined totals of testing and training sets) across the entire dataset was 87.41%.

At this time we are unaware of any other heparin-binding prediction models, and thus have no precedent to compare these results to.

6 Conclusions and future work

A heparin-binding prediction model is proposed. It is a Support Vector Machine based model using protein primary structure as input. The model considers XB pattern frequency features at the present time. A preliminary prototype system is developed that allows a user to provide an amino acid sequence, pick a model, and then the system returns the prediction result to the user.

The models and the software described in this paper have demonstrated significant potential in advancing research efforts of identifying new heparin- binding proteins via the investigation and study of protein and peptide sequences as a means of developing new, more effective treatment methods for many diseases. By using XB patterns as features, this investigation provides additional insight into the role that XB patterns or motifs play in proteins that bind to heparin.

To improve the prediction accuracy, we will investigate the selection of other XB patterns and features beyond XB patterns such as chemical and physical properties of amino acids. Since protein interaction and protein binding take place in 3-dimensional space, including secondary and tertiary structural information would likely improve the accuracy of the prediction model and will be studied.

Additionally, other types of machine learning techniques may yield better results to this type of problem. However, given the limited dataset available for binding and non-binding proteins, finding other suitable techniques may be difficult.

7 Acknowledgements

This research was supported by an Honors College International Research Grant.

We would like to thank Dr. Kumar and Dr. Srinivas Jayanthi of the Chemistry and Biochemistry department of the University of Arkansas for providing the protein data and guidance with the biological aspect of this research.

8 References

[1] Casu, B., Guerrini, M., and Torri. G. (2004) Structural and conformational aspects of the anticoagulant and anti-thrombotic activity of heparin and dermatan sulfate. Curr Pharm Des 10: 939-949.

[2] Gandhi N. S. and Mancera R. L. (2008) The Structure of Glycosaminoglycan and their interaction with Proteins, Chem Biol Drug Des 2008, 72:455-482

[3] C.-C. Chang and C.-J. Lin. (2011) LIBSVM A library for support vector machines, ACM Transactions on Intelligent Systems and Technology, 2:27:1-27:27.

[4] Shigeo Abe, Support Vector Machines for Pattern Classification, Second Edition, Springer, 2010

[5] C. Cortes and V. Vapnik. (1995) Support-vector network, Machine Learning, September 1995, Volume 20, Issue 3, pp273-297

[6] Arunkumar, A. I., Srisailam, S., Kumar, T. K., Kathir, K. M., Chi, Y. H., Wang, H. M., et al. (2002) Structure and stability of an acidic fibroblast growth factor from Notophthalmus viridescens. J Biol Chem, 277(48), 46424-46432

[7] Capila, I. and Linhardt, R. J. (2002) Angew. Chem. Int. Ed. Engl. 2002, 41, 391-412.

[8] Dempewolf, C., Morris J., Chopra M., Jayanthi S., Kumar T., and Li W. (2013) Identification of Consensus Glycosaminoglycan Binding Strings in Proteins Proc. International Conference on Information Science and Applications, 310-314.

[9] Arunkumar, A. I., Kumar, T. K., Kathir, K. M., Srisailam, S., Wang, H. M., Leena, P. S., et al. (2002) Oligomerization of acidic fibroblast growth factor is not a prerequisite for its cell proliferation activity. Protein Sci, 11(5), 1050-1061.

[10] Bae, J., Desai, U. R., Pervin, A., Caldwell, E. E., Weiler, J. M., and Linhardt, R. J. (1994) Interaction of heparin with synthetic antithrombin III peptide analogues. Biochem J, 301 (Pt 1), 121-129.

[11] Adjit Varki, Richard D Cummings, Jeffrey D Esko, Hudson H Freeze, Pamela Stanley, Carolyn R Bertozzi, Gerald W Hart, and Marilynn E Etzler, Essentials of Glycobiology, 2nd Edition, Cold Spring Harbor, 2009

[12] Hsu, C., Chang, C., and Lin, C. "A Practical Guide to Support Vector Classification" Internet: https://www.csie.ntu.edu.tw/~cjlin/papers/guide/guide.pdf, April 15th, 2010 [December 7, 2015].

[13] http://www.uniprot.org

[14] http://biopython.org/

[15] Chih-Chung Chang and Chih-Jen Lin, LIBSVM : a library for support vector machines. ACM Transactions on Intelligent Systems and Technology, 2:27:1--27:27, 2011. Software available at http://www.csie.ntu.edu.tw/~cjlin/libsvm

[16] Fang J, Dong Y, Salamat-Miller N, Middaugh CR. DB-PABP: a database of polyanion-binding proteins. Nucleic Acids Res. 2008 Jan; 36(Database issue):D303-6. Epub 2007 Oct. 4

Phosphatidylserine Torus as a Macromolecular Scaffold for the $GABA_A$ Receptor

S.F. Reader[1], J.L. Mustard[2], and N.W. Seidler[1,2]
[1]Department of Anesthesiology, University of Missouri-Kansas City School of Medicine, Kansas City, Missouri, United States of America
[2]Laboratory of Computational Biology and Structural Bioinformatics, Division of Basic Sciences, Kansas City University of Medicine and Biosciences, Kansas City, Missouri, United States of America

Abstract - *Understanding of the role of membrane lipids in the structure and function of neuronal receptors is currently incomplete. Annular lipids have been shown to exhibit an affinity to integral proteins by virtue of the hydrophobic complementarity of the lipid acyl chains and the protein amphipathic transmembrane helices. However, little is known about the role of the membrane lipid head-groups in maintaining affinity, or even structural integrity, at the lipid-protein interface. We used molecular modeling techniques to examine the interaction of phosphatidylserine and the $GABA_A$ receptor and propose that a torus of phosphatidylserine molecules acts as a scaffold for this receptor and facilitates higher order macromolecular complexes, involving cytosolic regulator proteins.*

Keywords: Phosphatidylserine, gamma-aminobutyric acid type A receptor, glyceraldehyde 3-phosphate dehydrogenase, biomembranes

1 Introduction

Lipids play a critical role in nervous system structure and function. Synaptic complexes and myelin exhibit unique compositions regulating specialized properties of these nervous system features. One of the major membrane phosphoglycerides, specifically phosphatidylserine, supports cognition in humans, including memory, learning, and higher order cognitive function [1]. Phosphatidylserine represents approximately 9% of total lipids in gray matter and 8% of total lipids in white matter [2]. Phosphatidylserine (Figure 1) is particularly enriched with the long chain polyunsaturated 22-carbon fatty acid, docosahexaenoic acid [3]. Approximately 42% of rat brain phosphatidylserine contains docosahexaenoic acid at the C-2 position. A recent publication on the role of phosphatidylserine in the human brain [1], which reviewed 127 Medline articles, identified this particular membrane phospholipid as crucial for neuronal and myelin health. We are interested in understanding the interaction of phosphatidylserine with a major receptor complex in brain, gamma-aminobutyric acid type A receptor ($GABA_AR$) (Figure 2), which regulates neuronal inhibition. Administration of phosphatidylserine was shown to modulate the cell content of $GABA_AR$ [4], as well as affect its function [5]. During purification of receptor (namely, detergent solubilization) addition of phosphatidylserine promotes stabilization of the protein complex. These observations suggest that phosphatidylserine not only interacts with the $GABA_AR$, but that its interaction may be crucial to maintaining its native structure.

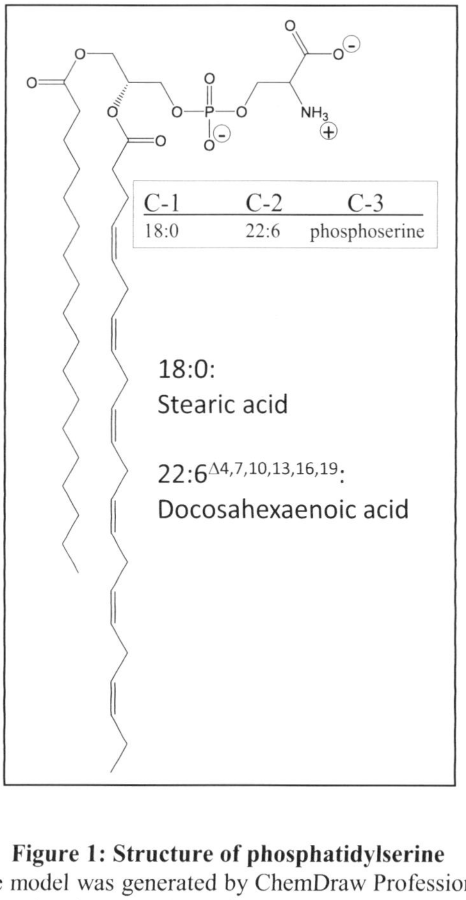

Figure 1: Structure of phosphatidylserine
The model was generated by ChemDraw Professional 15.0, showing the charged atoms on the headgroup (net charge of +1) and the two most-common fatty acids associated with this phosphoglyceride. In our molecular dynamics simulations, we constructed a phosphoserine mimetic to resemble the head group.

In biomembranes, most of the phospholipids represent bulk solvent in the bilayer. Others create a torus or annulus around an integral protein, such as a receptor. These phospholipids are thought to exhibit hydrophobic compatibility with the protein's transmembrane domain. The role of the phospholipid headgroup in forming the annular structure around integral proteins is poorly understood. We think that the phospholipid head groups may also play a role in creating a torus-like structure surrounding the transmembrane channel of the ionophoric receptor. We use the term 'torus' for this annular arrangement of phosphatidylserine lipids, since the C-2 docosahexaenoic acid hydrophobic chain exists in a curved spiral orientation.

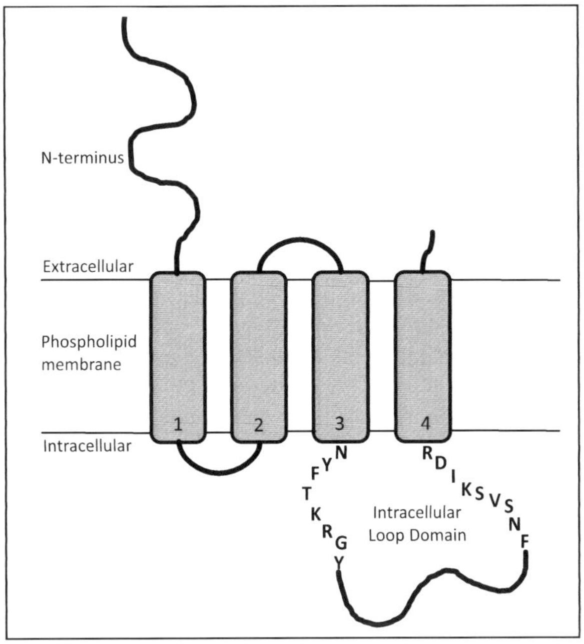

Figure 2: Schematic of $GABA_AR$
Illustration shows a single subunit of the pentameric $GABA_AR$ channel in neuronal membranes, presenting the large N-terminal domain, four single-pass transmembrane helices (TM 1-4), and the intracellular loop domain (ILD). The residues shown represent the *adjacent* regions (TM3A and TM4A), which exhibit significant unstructured stretches [6].

The purpose of the study is to examine a tripartite assembly of the receptor complex, including the pentameric $GABA_AR$ channel, the phosphatidylserine torus, and the intracellular regulator, known as GAPDH (glyceraldehyde 3-phosphate dehydrogenase).

GAPDH is a highly conserved multifunctional glycolytic enzyme. It is known to regulate the $GABA_A$ receptor [7], a target of anesthetics [8]. Isoflurane enhances GAPDH binding to the intracellular loop domain (ILD), the cytoplasmic chain that connects transmembrane helices 3 and 4 (TM3 and TM4, respectively) of the $GABA_AR$ [9]. The structural features of the ILD in $GABA_AR$ regulation are unknown as it was bioengineered out prior to recent crystal structure analysis [10]. We were interested in determining how the ILD interacts with both GAPDH and the cytoplasmic phospholipid membrane in order to elucidate its structure and function.

2 Methods

In our initial molecular dynamics studies involving individual molecules of phosphatidylserine with ILD peptides, we encountered issues with the promiscuity of the acyl chains and as such we constructed head group mimetic attached to propanol to represent phosphoserine moiety.

2.1 Protein sources

Human GAPDH crystal structure was obtained from 1U8F.pdb [11]. The existing structure is an asymmetric homo-tetramer with subunits designated as O, P, Q, and R. This molecule was uploaded directly into ChemBio3D 15.0 (cambdrigesoft.com) and then modified to include only the P and Q subunits, and their bound coenzymes (NAD^+), forming the major groove (Figure 3). The PQ dimer is stabilized across the *R*-axis, which involve a substantial number of contacts that are considered highly conserved [12]. The O and R subunits, as well as the single remaining NAD^+, were deleted to simplify molecular dynamics computations. All water molecules within 4Å of the protein surface were kept for molecular dynamics simulations.

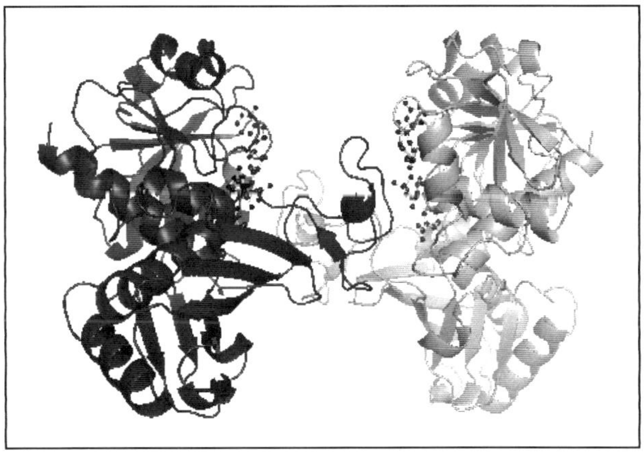

Figure 3: Diagram of the PQ dimer of GAPDH
Illustration shows the P (*dark ribbon*) and Q (*grey ribbon*) subunits with the bound NAD+ coenzymes (*black ball & stick*). Obtained from ncbi/structure and modified in PyMOL Molecular Visualization. The major groove is the upper cleft flanked by the lobes of the P and Q subunits. The lower cleft is normally occupied by the other OR dimer that makes up the native tetrameric GAPDH.

$GABA_AR$ structure was obtained from 4COF.pdb [10]. It is a homopentameric structure consisting of five β3 subunits. This molecule was also directly uploaded into ChemBio3D and then modified to include only one of the β3 subunits. It is important to note this subunit lacked the entire ILD as it was selectively deleted to obtain the crystal structure.

2.2 Creation of ILD peptides

Peptide sequences of the ILD adjacent to TM3 and TM4 were created in ChemDraw Professional 15.0 using amino acid sequences found on uniprot.org (p14867, α1 subunit). The peptide sequence, designated TM3A, consisted of 10 residues, `TVNYFTKRGY`, where `TV` is part of TM3 and `NYFTKRGY` represents the initial ILD sequence. The other peptide sequence, designated TM4A, consisted of 11 residues, `FNSVSKIDRLS`, where `LS` is part of TM4 and `FNSVSKIDR` represents the end of the ILD sequence. An additional residue for TM4A was included as it provided stability necessary to undergo the molecular dynamics simulations.

2.3 Creation of phosphoserine compounds

Phosphoserine (PS) containing molecules, representing the phospholipid membrane, were also created: di-stearoyl phosphatidylserine (DSPS) and 1-propanol phosphoserine (PPS). DSPS was first created to model the phospholipid membrane. PPS was next created to avoid the interfering interactions of the large acyl chains of DSPS. The structure of a typical biomembrane would normally prevent such interaction with the large hydrophobic acyl chains. PPS consisted of a glycerol backbone with a phosphoserine attached to C1 of glycerol and removed hydroxyls from C2 and C3, creating a propanol backbone.

2.4 Molecular dynamics

We used computer-based molecular modeling, ChemBio3D, to assess the interaction of the ILD and its two binding partners: GAPDH and the phosphatidylserine head group. Molecular dynamics (MD) simulations were performed under explicit and implicit (i.e. no water molecules) conditions. All MD simulations were begun with MM2 energy minimization of each molecule, so that the cumulative potential energy was minimized and areas of strain were corrected. Prior to simulations involving more than one molecule, each molecule was minimized separately first and then again when combined. An explanation of the algorithm that was used in MM2 force field computations can be obtained from the vendor (cambridgesoft.com). The molecular dynamics simulations were undertaken after energy minimization with a heating rate of 1kcal/atom/ps selected with a target temperature of 300° Kelvin. The step and frame intervals were 2.0fs and 10fs, respectively; and the simulation was set for termination after 10,000 iterations. Following MD simulations, intermolecular distances and dihedral angles were measured as described below.

2.5 Intermolecular distances and dihedral angle measurements

Following MD simulations between TM3A (or TM4A) with PPS (or DSPS), intermolecular distances (*t*-test using $p<0.05$) and peptide main chain dihedral angles were measured and compared. Intermolecular distances were obtained between the phosphate atoms of PPS (or DSPS) and the basic (nitrogen atoms of arginine or lysine) and neutral residues. Residues arginine-340 and lysine-339 from TM3A were used. Arginine-423 and lysine-420 from TM4A were used. The neutral residues were selected so that they were distributed evenly throughout the peptide, indicating no preference for either N- or C-terminus. For TM3A, the oxygen atoms of neutral residues asparagine-335, tyrosine-336, threonine-338 and tyrosine-342 were used. For TM4A, the oxygen atoms of neutral residues asparagine-416, serine-417, serine-419, and serine-425 were used. Twenty simulations were performed with TM3A in the presence of PPS and eighteen with DSPS. The number of simulations involving TM4A was less due to the issues described below. We also measured main chain dihedral angles surrounding the alpha carbons of TM3A and TM4A following MD simulations pre and post interaction with PPS. Ramachandran diagrams of the phi and psi dihedral angles were prepared and charts were overlaid for comparison. We chose an minimal threshold value of 90° to indicate a significant change in the phi or psi dihedral angles.

For the MD simulations between TM3A (or TM4A) with the GAPDH PQ dimer, the molecules were initially positioned 10 to 14Å apart in random positions and proximity after simulation was measured indicating the GAPDH residues that were closest. Multiple succeeding simulations were performed in most cases in order to reach an end point of closer proximity. Seven distinct trials were initiated with TM3A that continued for a total of 18 simulations. There were five distinct trials for TM4A that continued for a total of 14 simulations.

2.6 Inter-atomic distances between TM3 and TM4

The average distance between TM3 and TM4 in the β3 subunit of $GABA_AR$ was also determined. These measurements were completed using an uploaded crystal structure (4COF.pdb) that did not undergo any prior MD simulation. The upper two-thirds of the neighboring helices were used to determine the inter-helical distance. We avoided the bottom one-third due to the bioengineered changes that occurred upon creation of the crystallized form of the $GABA_AR$, which omitted the ILD [10]. Two β3 subunits, A and C, were used; alpha carbons of multiple pairs of residues were selected and their inter-atomic distances were measured. The positioning of the residue pairs was chosen based on visual inspection to evenly distribute and align the pairs of residues along the longitudinal axes of the two helices. Average values were then obtained from the 10 total inter-helical distances obtained. It was inferred that the distances between TM3A and TM4A, which emanate immediately from TM3 and TM4, exhibit approximately the same inter-chain distances. In order to validate the model of GAPDH-GABA$_A$R ILD interaction, we also analyzed the topology of the GAPDH PQ dimer's major groove. The distance across the major groove formed by the PQ subunits was measured using an uploaded crystal structure (1U8F.pdb) that did not undergo any prior MD simulation. The alpha carbons of residue arginine-80 from both P and Q subunits at the opposing edges of the major groove were selected and their inter-atomic distance determined. These residues were chosen due to their location, representing the widest-most region of the major groove that would facilitate occupancy of the ILD. Curiously, arginine-80 is also part of the GAPDH-phosphatidylserine binding site (residues 72-96) [13].

3 Results

3.1 Conformational change of ILD triggered by phospholipid

In a comparison of TM3A dihedral angles with and without PPS following MD simulations, we observed that PPS promoted a conformational change (Figure 4) in this region of the ILD. Upon visual inspection, several of the side chains appeared to be repositioned, suggesting that the conformation of the ILD is in part influenced by the phospholipid membrane.

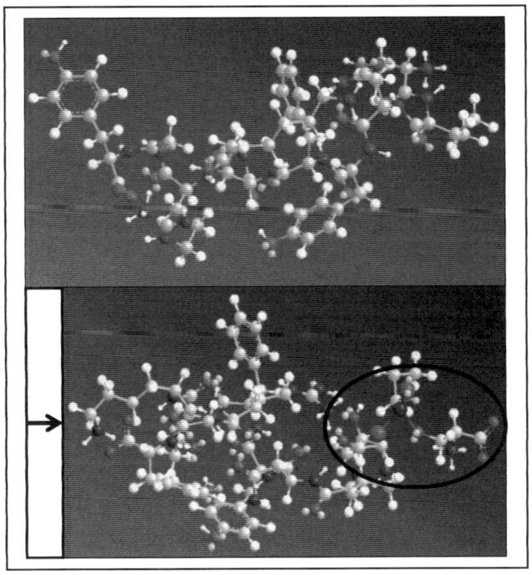

Figure 4: Interaction between TM3A and PPS
TM3A after molecular dynamics alone (*top*) and with PPS (*bottom*). Top image demonstrates a more extended conformation, compared to bottom image, which contains PPS (*black oval*).

While TM3A has a propensity for an α-helical structure, it is neighbored by a peptide region downstream that is highly disordered [6]. The presence of the phospholipid membrane may stabilize the TM3A region in a preferred conformation.

The phi and psi dihedral angles around five residues of TM3A changed as a result of interaction with PPS to a greater than 90° shift (Figure 5). The dashed arrows in the Ramachandran diagram indicate the large changes in position of the following residues: valine-334, asparagine-335, phenylalanine-337, threonine-338, and lysine-339. This observation suggests that the membrane-dependent repositioning of threonine-338 may facilitate phosphorylation of this residue, which we know regulates GABA$_A$R chloride channel activity [7]. This observation is consistent with the finding that phospatidylserine plays a role in modulation of the receptor [4].

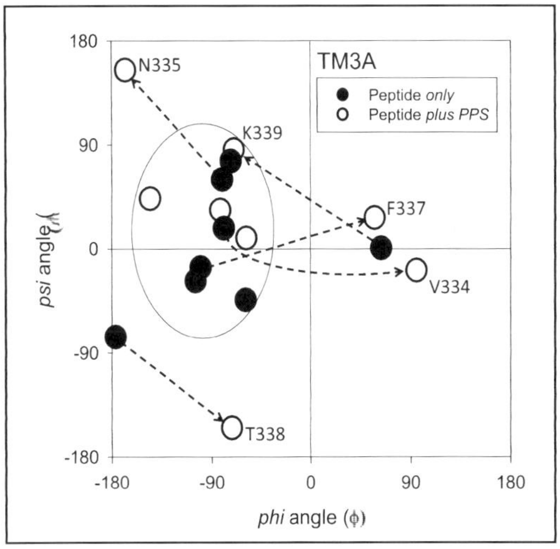

Figure 5: Dihedral angles of TM3A
Comparison of TM3A dihedral angles alone (*filled symbols*) and after (*open symbols*) interaction with PPS. Several of these angles changed in the presence of PPS, suggesting a rather large change in conformation.

In the comparison of TM4A dihedral angles with and without PPS following MD simulations, we again observed that PPS promoted a conformational change in the C-terminal region of the ILD. Most interactions between TM4A and PPS exhibited repulsion in our MD simulations, with one exception, in which PPS exhibited a specific orientation imbedded in the core of the peptide. The phi and psi dihedral angles around three residues of TM4A changed as a result of interaction with PPS to greater than a 90° shift (Figure 6). The dashed arrows in the Ramachandran diagram indicate the large changes in position of the following residues: serine-417, aspartate-420, and arginine-421. An additional residue, asparagine-414, shifted 88°. The phosphorylation site on this area of the ILD is serine-417 [7], suggesting again that the membrane-dependent repositioning of this target residue may facilitate phosphorylation.

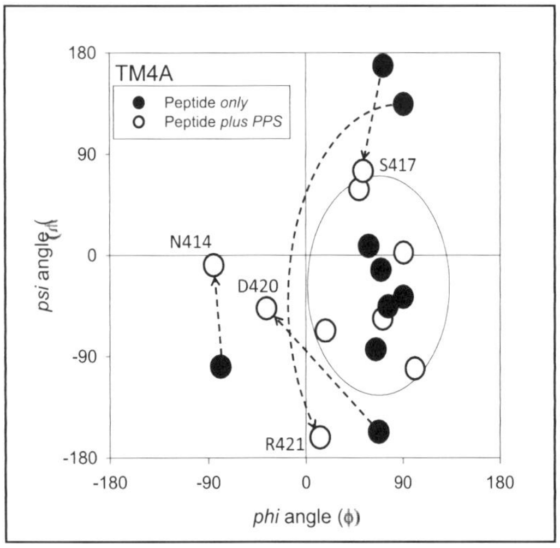

Figure 6: Dihedral angles of TM4A
Comparison of TM4A dihedral angles alone (*filled symbols*) and after (*open symbols*) interaction with PPS. Several of these angles changed in the presence of PPS, suggesting a rather large change in conformation.

The use of our phospholipid-mimetic, PPS, was successful and enabled to assess ILD-phospholipid head group interaction. Future implementation of this chemical in MD simulations may offer researchers a useful tool for studying phosphatidylserine interaction with cytosolic compounds without the interference of acyl chains.

3.2 Specificity of ILD and Phospholipid Interaction

In response to the above observations regarding the PPS-induced conformational change of TM3A, we were interested to further the understanding of the nature of this interaction. Following MD simulations as described in Methods, we measured the distances of the basic, versus the neutral, residues on TM3A to the phosphate atoms on the phospholipid-mimetics. We observed that the basic residues trended towards a closer proximity to the phosphate atom on PPS compared to the neutral residues with a near significant *p*-value using the 95% confidence limit ($p = 0.051$). In MD simulations using TM3A and DSPS, we observed again that the basic residues trended towards a closer proximity to the phosphate atom than neutral residues, reaching significance ($p < 0.01$). These observations suggest the importance of the basic residues in ILD-phospholipid membrane interaction. The TM3A lysine and arginine residues at this specific location are highly conserved across multiple families of pentameric ligand gated ion channels [14]. O'Toole and Jenkins [14] observed that two basic residues were crucial in understanding $GABA_AR$ function. Mutation of these residues impaired receptor activity. Our observations suggest that these residues may play a role in organizing the structure of the ILD, in relation to the phospholipid membrane.

3.3 Specific interaction of ILD at the major groove of the GAPDH

We were interested in determining whether the ILD regions, existing directly at the membrane surface, also displayed specific interactions with the $GABA_AR$ regulator, GAPDH. The consensus sites for GAPDH mediated phosphorylation of the ILD are contained in the TM3A and TM4A sequences [7]. This literature observation implies that GAPDH binds and interacts with these ILD regions. We tested this idea by performing MD simulations involving TM3A (and TM4A) with GAPDH. We observed that TM3A reproducibly interacted with the GAPDH PQ dimer at a similar location at the opposing edges of the major groove, therefore involving both subunits (Figure 7).

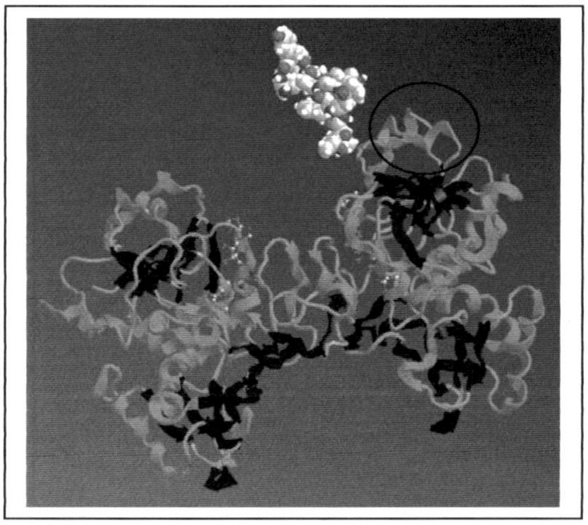

Figure 7: Interaction of TM3A and GAPDH
Tracking the interaction of TM3A and GAPDH PQ dimer using MD simulation. TM3A (*space-filling model*) is shown interacting at major groove of GAPDH PQ dimer (*ribbon diagram*) to within 4Å following three different simulations. The phosphatidylserine binding sites (residues 72-96 on each lobe of GAPDH), one of which is shown in *black oval*. For image clarification, water molecules were removed after simulations.

Of the seven trials of MD simulations performed, four were successful in going from an intermolecular starting distance of 12Å to an endpoint of 4Å. We identified a region (residues 78-84), which was found to be within 4Å of TM3A following these successful MD simulations. The residues on GAPDH that represented the most reproducible points of contact with TM3A (three of the four successful trials) were arginine-80, aspartate-81, proline-82, and lysine-84. Additionally, three residues (glutamine-78, glutamate-79, serine-83) were found to be within 4Å following two of the four trials. Curiously, this region is in the same location as the phosphatidylserine binding site of GAPDH (residues 72-96) [13]. The identical pocket on the cytosolic GAPDH contains binding sites for the ILD *as well as* phosphatidylserine; this strongly suggests that GAPDH may form a dynamic dual docking site at the surface of the inner membrane.

3.4 Distance measurements in support of macromolecular working model

To assess the feasibility of our model, we measured the distance between $GABA_AR$ TM3 and TM4 (which we determined to be an average of 13Å using 1U8F.pdb), as well as the GAPDH major groove (which we determined to be and average of 22Å using 4COF.pdb), suggesting that the ILD is able to fit in the groove. Given the presence of the phosphatidylserine-binding sites at the tips of the lobes of the PQ dimer, we propose that the GAPDH dimer straddles the TM3-TM4 segments at the point of entry into the cytosolic space.

4 Discussion

We studied the interactions of the ILD and membrane components. The ILD sequences directly adjacent to TM3 and TM4 exhibited a large conformational change upon phospholipid interaction (measured by peptide chain dihedral angles). This ILD-PS interaction also demonstrated specificity (evidenced by the closer proximity of basic compared to neutral residues). In addition, the ILD peptides interacted at the edge of the GAPDH major groove, which also contains the phosphatidylserine binding site, allowing for a dynamic tripartite interaction. Based on these results, we developed a working model of these interactions (Figure 8).

We propose that GAPDH dynamically interacts with the phosphatidylserines directly surrounding the $GABA_AR$. Concurrently, GAPDH straddles the ILD within the major groove. This is consistent with the literature as isoflurane enhances GAPDH-ILD interaction [9]. This tripartite arrangement may be influenced by the ILD-PS interaction. These observations provide insight into the structural features of the $GABA_AR$ ILD, which previously was unknown [10].

5 Conclusions

Our study identified specific interactions involving the $GABA_AR$, the surrounding membrane, and GAPDH. Our proposed macromolecular complex may be modulated by isoflurane, availability of phosphatidylserine, and the structural integrity of GAPDH.

Our findings suggest that the surrounding phosphatidylserine lipids act as a molecular scaffold that can offer interactions with GAPDH and $GABA_AR$. One can envision that the GAPDH binding pocket can accommodate both the TM3A ILD and the phosphatidylserine head group. Conversely, this docking site may involve only one (i.e. either ILD or phosphatidylserine), making the GAPDH interaction more dynamic.

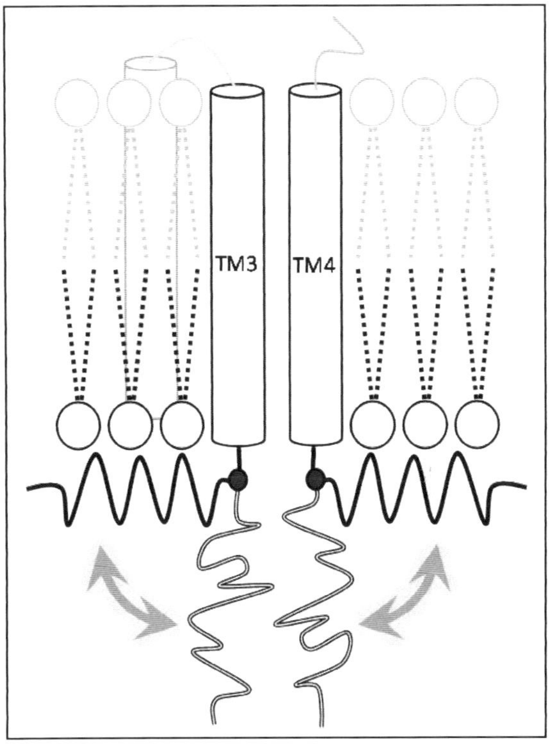

Figure 8: Model of phosphatidylserine torus
The transmembrane helices of $GABA_AR$ are given as *cylinders*. The surrounding phosphatidylserine molecules are illustrated as *circles* (for head groups) and *dotted lines* (for acyl chains). The arrows indicate the dynamic movement of protein segments (TM3A and TM4A) to the scaffold provided by the phosphatidylserine torus. Once in place at the inner membrane surface, GAPDH can straddle the TM3-TM4 doublet stabilizing the macromolecular complex.

The use of our phospholipid-mimetic, PPS, was successful and enabled to assess ILD-phospholipid head group interaction. Future implementation of this chemical in MD simulations may offer researchers a useful tool for studying phosphatidylserine interaction with cytosolic compounds without the interference of acyl chains. We think that the annular ring of phosphatidylserine molecules resemble more like a torus due to the spiral nature of docosahexaenoic acyl chains.

6 References

[1] Glade MJ, Smith K. Phosphatidylserine and the human brain. Nutrition 2015;31(6):781-6.

[2] Suzuki K (1981) Chemistry and metabolism of brain lipids. In G.J. Siegel, R.W. Albers, B.W. Agranoff & Katzman R (Eds.) basic Neurochemistry (pp. 355-370) (3rd Ed.) Boston, MA: Little, Brown & Co.

[3] Lee CH, Hajra AK.Molecular species of diacylglycerols and phosphoglycerides and the postmortem changes in the molecular species of diacylglycerols in rat brains. J Neurochem. 1991;56(2):370-9.

[4] Levi de Stein M1, Medina JH, De Robertis E. In vivo and in vitro modulation of central type benzodiazepine receptors by phosphatidylserine. Brain Res Mol Brain Res. 1989;5(1):9-15.

[5] Hammond JR, Martin IL. Modulation of [3H]flunitrazepam binding to rat cerebellar benzodiazepine receptors by phosphatidylserine. Eur J Pharmacol. 1987 May 7;137(1):49-58.

[6] Mustard JL, Worley JB, and Seidler NW. Architectural Topography of the α-Subunit Cytoplasmic Loop in the $GABA_A$ Receptor. IN: (Q-N Tran, HR Arabnia, editors) "Emerging Trends in Computational Biology, Bioinformatics, and Systems Biology: Systems and Application" within book series, Emerging Trends in Computer Science and Applied Computing", Chapter 7 pages 19-33, Elsevier 2016.

[7] Laschet JJ et al. Glyceraldehyde-3-phosphate dehydrogenase is a $GABA_A$ receptor kinase linking glycolysis to neuronal inhibition. J Neurosci 2004;24(35):7614-22.

[8] Schofield CM, Harrison NL. "Transmembrane residues define the action of isoflurane at the GABAA receptor alpha-3 subunit." Brain Research, 2005;1032(1-2):30-35.

[9] Montalbano AJ, Theisen CS, Fibuch EE, Seidler NW. Isoflurane enhances the moonlighting activity of GAPDH: implications for $GABA_A$ receptor trafficking, ISRN Anesthesiology, Volume 2012 (2012), Article ID 970795, 7 pages.

[10] Miller PS, Aricescu AR. "Crystal structure of a human $GABA_A$ receptor." Nature, 2014; 512 (7514): 270-275.

[11] Jenkins JL and Tanner JJ. High-resolution structure of human d-glyceraldehyde-3-phosphate dehydrogenase. *Acta Crystallogr., Sect D.* 2006;62:290

[12] Seidler NW. Dynamic oligomeric properties. *Adv Exp Med Biol.* 2013;985:207-247

[13] Kaneda M, Takeuchi K, Inoue K, Umeda M. Localization of the phosphatidylserine-binding site of glyceraldehyde-3-phosphate dehydrogenase responsible for membrane fusion. J Biochem. 1997;122(6):1233-40.

[14] O'Toole KK, Jenkins A. Discrete M3-M4 intracellular loop subdomains control specific aspects of γ-aminobutyric acid type A receptor function. J Biol Chem. 2011;286(44):37990-9

SESSION

GENE EXPRESSION, REGULATORY NETWORKS, MICROARRAY, SEQUENCING, ALIGNMENT, AND RELATED STUDIES

Chair(s)

TBA

An Anomaly Detection Algorithm for Identifying Alien Gene Clusters in Microbial Genomes

Dongsheng Che[1], **Sai Vahini Manikonda**[1], **Zuqing Li**[1], **and Bernard Chen**[2]
[1]Department of Computer Science, East Stroudsburg University, East Stroudsburg, PA 18301, USA
[2]Computer Science Department, University of Central Arkansas, Conway, AR 72034, USA

Abstract— *Genetic materials are often transferred between microbial organisms. The host genome regions that contain alien genetic materials may carry integration-related genes and products in some cases, but not all the time. Therefore, computational methods that use such information for detecting alien gene clusters in genomes have some limitation. The genome composition based approached have been applied using oligonucleotides and codon usage. In this paper, we report the development of our parametric genome composition-based anomaly detection algorithm for alien gene cluster finding, using the parameters of genome block size and k-mer word. We tested our algorithm on five genomic sequences. Our prediction results have shown that our algorithm can accurately detect alien gene clusters in these genomes.*

Keywords: Alien gene cluster; Anomaly detection; Archaea; Bacteria; Gaussian distribution; Genome sequence

1. Introduction

An outlier is the observation that is significantly different than the majority of the observations, and there are many applications for outlier detection, such as fraud detection, and intrusion detection [1]. In bacterial and archaeal genomes, outlying genome segments are often seen. In this case, DNA patterns in the outlier regions are distinct from the remainder of the host genomes. Typically, two lineage distant organisms should have different genome patterns since each organism genome pattern is unique. Therefore, a genome with outlying regions in one organism could be caused by the gene inflow from other organisms, which is also known as *horizontal gene transfer* (HGT).

The gene inflow from the alien organism to the host organism can be carried out through: transformation [2], conjugation [3], transduction [4], and gene transfer agent [5]. During the integration of the alien genome sequences into the host genomes, integrase, transposon genes might participate in the integrating process in some cases [6]. The alien genome regions are sometimes also flanked by transfer RNAs [7].

The identification of alien gene clusters in the host genome is extremely important for biological communities. It helps evolutionary biologists understand how bacterial genomes are evolving in organisms' evolutionary history. Such findings could also be used for studying donor-recipient relationship through gene cluster transfers, and for constructing donor-recipient networks across genomes [8]. In other cases, where alien gene clusters contain pathogenic genes, these can be used for explaining why some non-pathogenic bacterial genomes were found to cause diseases [9].

Alien gene clusters were initially found in *Escherichia coli* [10]. Later on, more biological experiments and evolutionary studies showed the existence in other bacaterial and arachaea genomes [11]–[14]. With the explosive growth of completed genome sequences, computational tools have been developed for predicting alien gene clusters. we can roughly categorize computational tools into three groups: sequence composition based, comparative genome analysis based, and ensemble based.

The sequence composition based approaches find the alien gene clusters based on the DNA signatures (such as G + C content, dinucleotide pattern), codon usage patterns, or integration associated genes (such as transposon genes and integrase genes, tRNA genes) in the query genome. The tools using sequence and gene information are AlienHunter [15], Centroid [16], GIDetector [17], GIHunter [18], PAI-IDA [19], and SIGI-HMM [20].

For the comparative genome analysis based approach, the identification of alien gene clusters rely on the reference genomes, which are lineage closely related to the query genome. The gene clusters, that only exist in the query genomes but not in its reference genomes, are considered to be alien gene clusters [6]. Representative tools for this group of approaches include IslandPick [21] and MobilomeFINDER [22]. Ensemble-based approaches integrate the prediction results of standalone tools to obtain census GIs. For instance EGID [23] integerates the prediction of AlienHunter [15], SIGI-HMM [20], INDeGenIUS [24], IslandPath [25], and PAI-IDA [19].

Despite the development of computational tools for detecting alien gene clusters, none of them can accurately predicts for all genomes. Some prediction tools have high sensitivity but with low specificity such as AlienHunter [15], while other have high specificity but with low sensitivity such as IslandPath [25]. Some other tools that incorporates mobile gene, tRNA gene or virus gene information in the models also suffer the prediction accuracy, due to the fact that all genomes having alien gene clusters have such characteristics

[26]. Therefore, there is still room for improvement in the detection of alien gene clusters.

In this study, we will not consider genomic specific feature information such as mobile genes, tRANs genes, or virus related genes for alien gene cluster prediction. Instead, we will focus on the DNA sequence only to make our prediction tool applicable to all genomes. Specifically, we will first build DNA based features that rely on a parametric block size (B) of the genome sequence. The DNA features are built through counting the frequencies of k-mer nucleotides. The anomaly detection algorithm will be implemented by evaluating all combined feature values based on Gaussian distribution model. The performance of the algorithm will be evaluated by applying our algorithm on completely sequenced genome datasets.

The remainder of this paper is organized as follows. Section 2 describes the anomaly detection algorithm. Section 3 shows the prediction results of our detection algorithm on five genomes. In Section 4, we will conclude the paper, with the discussion of future work.

2. Methods

We detect outlying genomic regions through analyzing each block of genomic segments. The block size (denoted as B) can range from 5 kb, 10 kb, 20 kb and 50 kb. The starting positions of each block are the multiple of $B/2$. For instance, for the block size of 20 kb, we look into the regions of (0-20), (10-30), (20-40) kb, etc. For each block of genomic region, we extract the features as described below:

2.1 Feature Construction

A vector of the features for any block of genomic region is constructed based on the k-mer words, where k can be 4, 5, or 6. Since there are four possible nuleotides (A, C, G and T) for any position of the k-mer word, the total number of possible words should be 4^k. For instance, for $k = 5$, then there are $4^5 = 1024$ different possible words, and thus the feature vector size is 1024. Since our block size of genomic region to be investigated are between $5k$ to $50k$, we choose the our k be 4, 5, and 6. If k is too large, then the frequencies of many of k-mer will be zeros, and thus making feature values indistinguishable in the dataset.

Let i be the ith block of genomic region, and m be the total number of blocks in the entire genome sequence. Let $x_j^{(i)}$ be the frequency of the jth word in the ith block of genomic region, then a vector of feature values $x^{(i)}$ in the ith block can be represented as follows:

$$x^{(i)} = (x_1^{(i)}, x_2^{(i)}, x_3^{(i)}, ..., x_{4^k-1}^{(i)}, x_{4^k}^{(i)}) \quad (1)$$

All m feature vectors ($x^{(1)}$, $x^{(2)}$,... and $x^{(m)}$ are computed, and they are used to be fed into the anomaly detection algorithm.

2.2 Anomaly Detection Algorithm

The outliers can be detected based on Gaussian distribution model. If a data point is far from the mean by more than a couple of times the standard deviation, then it can be treated as an outlier. For the problem of detecting outlying genomic region, we compute a vector of 4^k features of means and standard deviation as follows:

$$\mu_j = \frac{1}{m}\sum_{i=1}^{m} x_j^{(i)} \quad (2)$$

$$\sigma_j^2 = \frac{1}{m}\sum_{i=1}^{m} (x_j^{(i)} - \mu_j)^2 \quad (3)$$

The probability of a vector of 4^k features is the simply the product of all individual probabilities, which are calculated based on Gaussian distribution. The product of 4^k probabilities will be very small, and could lead to the problem of underflow error. To deal with this kind of problem, we could do log transformation. So the log of the product becomes the sum of logs. The probability for estimating the ith block of genomic region to be outlying or not can be given as follows:

$$LogP(x^{(i)}) = \sum_{i=1}^{4^k} log(\frac{1}{\sqrt{2\pi}\sigma_j} exp(-\frac{(x_j^{(i)} - \mu_j)^2}{2\sigma_j^2})) \quad (4)$$

The ith block of region is considered to be an outlying regions if $LogP(x^{(i)})$ is less than a predefined threshold value. The threshold value will be different for a different k-mer word. The pseudocode of the anomaly detection algorithm is given below:

Algorithm 1 Anomaly Detection Algorithm

```
m = the total number of blocks of genomic regions
for i ← 1, m do
    Compute the i^th block's feature vector x^(i)
    x^(i) = (x_1^(i), x_2^(i), x_3^(i), ..., x_{4^k-1}^(i), x_{4^k}^(i))
end for
for j ← 1, 4^k do
    μ_j = (1/m) Σ_{i=1}^{m} x_j^(i)
    σ_j^2 = (1/m) Σ_{i=1}^{m} (x_j^(i) − μ_j)^2
end for
for i ← 1, m do
    LogP(x^i) = Σ_{i=1}^{4^k} log( 1/(√(2π) σ_j) exp(−(x_j^(i) − μ_j)^2 / (2σ_j^2)) )
    if LogP(x^i) < threshold then
        Report i^th block to be an outlying genomic region
    end if
end for
```

Table 1: A list of genome datasets tested in this study

Genome	Length (Mb)	Accession Number	Reference
Corynebacterium glutamicum ATCC 13032	3.28	NC_003450.3	[27]
Corynebacterium diphtheriae NCTC 13129 chromosome	2.50	NC_002935	[28]
H. pylori strain J99	1.65	NC_000921	[29]
Rhodopseudomonas palustris CGA009	5.46	NC_005296.1	[30]
Vibrio vulnificus CMCP6 chromosome I	3.25	NC_004459.3	[31]

3. Results

We have implemented our parametric anomaly detection algorithm for alien gene cluster finding, and tested the algorithm on five genomes. The genome sequences were collected from the National Center for Biotechnology Information (NCBI) FTP server (ftp://ftp.ncbi.nih.gov/genomes/Bacteria). These five genomes were *Corynebacterium glutamicum* (*C. glutamicum*) ATCC 13032, *Corynebacterium diphtheriae* (*C. diphtheriae*) NCTC 13129 chromosome, *Helicobacter pylori* (*H. pylori*) strain J99, *Rhodopseudomonas palustris* (*R. palustris*) CGA009, *Vibrio vulnificus* (*V. vulnificus*) CMCP6 chromosome I. The genome information is shown in Table 1.

We run our algorithm using the following nine parameter combinations, ($5K$, 4-mer), ($10K$, 4-mer), ($20K$, 4-mer), ($50K$, 4-mer), ($10K$, 5-mer), ($20K$, 5-mer), ($50K$, 5-mer), ($20K$, 6-mer) and ($50K$, 6-mer). Below we present the predicted alien gene clusters, and compare the predicted results with the results using other methods.

3.1 *C. glutamicum* ATCC 13032

C. glutamicum ATCC 13032 is bacterium that has a genome length of 3.28 Mb, and it contains 3002 protein-coding genes. The species is important for industrial production of amino acids [32], and its genome contains several genome regions that were acquired from *C. diphtheriae* through horizontal gene transfer.

The prediction results of our algorithm with nine combinations of two parameters are shown in Figure 1. In general, the smaller block size and the smaller k-mer word, the more predicted outlying regions. On the other hand, the larger block size and the larger k-mer word, the less predicted outlying regions. For instance, the combination of ($5K$, 4-mer) in Figure 1(a) shows many valleys, while the combination of ($50K$, 6-mer) in Figure 1(i) shows only one significant valley. This significant outlying region of 1.75-2.00 Mb covers 236 genes, which was consistent with previous study of 1.78-1.99 Mb. The predicted region is mainly composed of hypothetic proteins, but it also contains one integrase gene, one phage related genes, seven tRNAs genes, and two transposases [26].

For the combination of ($20K$, 6-mer), as shown in Figure 1(h), our algorithm predicted four significant regions with the probablity value less than -4600, *i.e.*, 0.37-0.39, 1.78-1.94, 2.70-2.72, 3.16 -3.18. The genomic region of 0.37-0.39 contains a number of proteins related to lipopolysaccharide synthesis sugar transferase, glycosyltransferase, which explains its impact on the production of L-aspartate-derived amino acids and vitamins.

3.2 *C. diphtheriae* NCTC 13129 chromosome

C. diphtheria NCTC13129 is a gram-positive bacterium that produces diphtheria toxin (DT), which causes the symptoms of diphtheria [28]. The genome sequence has the size of 2.49 Mb, and it contains 2320 genes.

With the combination of ($50K$, 6-mer), we predicted four genomic regions that have the probability values less than -5200, *i.e.*, 0.15-0.20, 0.25-0.30, 0.35-0.40 and 1.10-1.40 (Figure 2(a)). The region of 1.12-1.52 contains 351 genes, including nine tRNA genes, and four putative transposase genes. this predicted region was consistent with other prediction tools such GIHunter [26] and IslandViewer [33].

3.3 *H. pylori* strain J99

H. pylori is one of the most common human pathogens that colonizes the gastric mucosa [29]. The genome is 1.65 Mb long, and it was reported to contain the type IV secretion system for virulence of pathogens, and encoded on the cag pathogenicity island, which is acquired by horizontal transfer [34].

For the combination of ($50K$, 5-mer), we predicted two outlying regions that have the probability values less than -1800, *i.e.*, 0.50-0.55, and 1.00-1.10 (Figure 2(b)). The region of 0.50-0.55 Mb contains 49 genes, with the majority of encoding cag island proteins. The region of 1.00-1.10 Mb contains 83 genes, with the majority of them encoding putative proteins, and with several integrases.

3.4 *R. palustris* CGA009

R. palustris CGA009 is a metabolically versatile photosynthetic bacterium that produces energy by using light and inorganic and organic compounds [30]. The genome is 5.46 Mb long, and it has 4,836 predicted genes. The

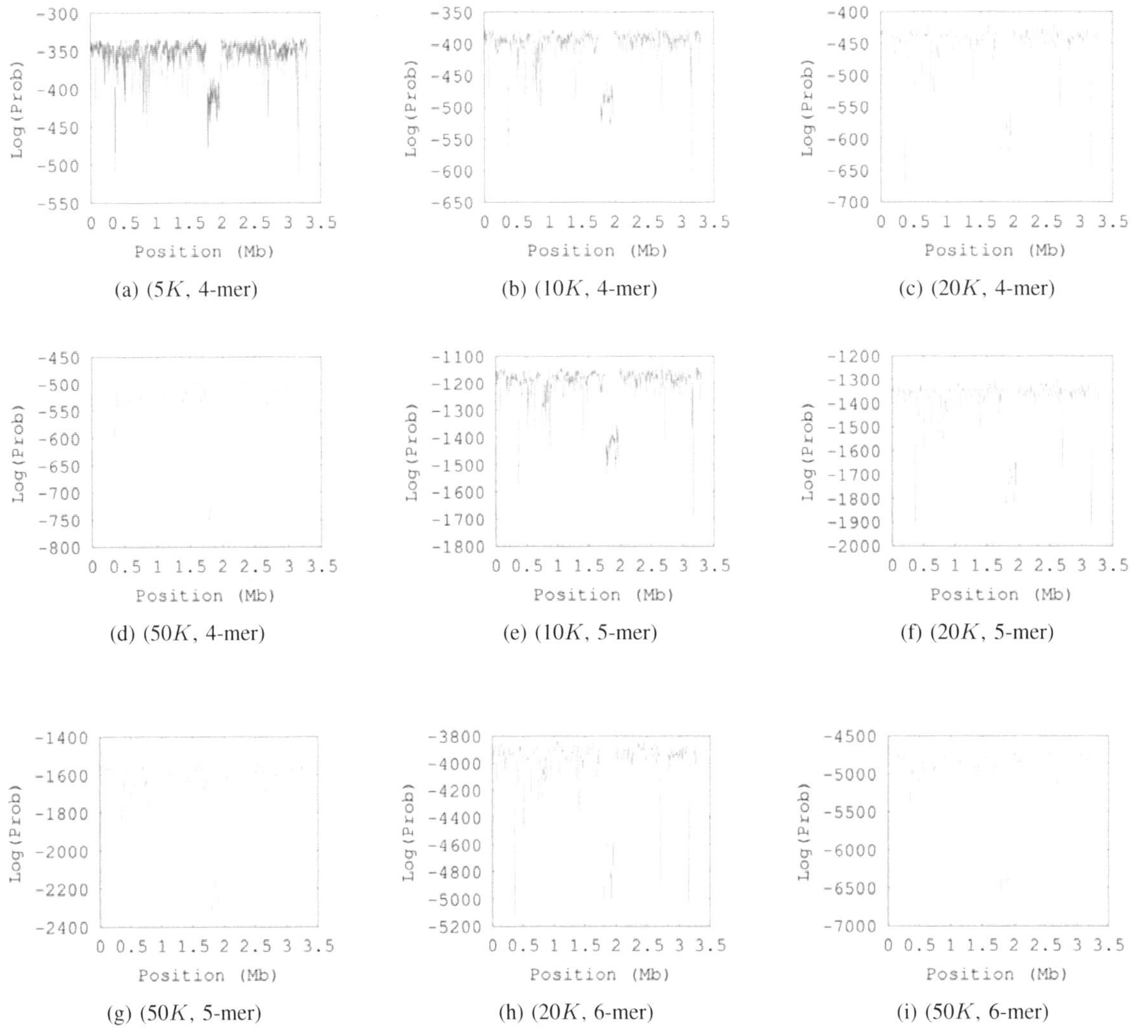

Fig. 1: Graphic representations for the probabilities in the genome of *C. glutamicum* ATCC 13032

G + C content did not show apparent horizontal transferr islands, but Z-curved based method was able to identify three horizontal islands [35].

With the combination of (50K, 6-mer), we predicted three genomic regions that have the probability values less than -5000, *i.e.*, 2.40-2.60, 3.70-3.80 and 4.55-4.70 (Figure 2(c)). The region of 2.40-2.60 contains 160 genes, including type IV secretion system subunit and conjugal transfer proteins. The region of 4.55-4.70 Mb contains multidrug-efflux transport proteins, conjugal transfer proteins, and other integration supporting integrases and tRNAs. All these three predicted regions were consistent with the reported regions with Z-curve approach [35].

3.5 *V. vulnificus* CMCP6 chromosome I

V. *vulnificus* CMCP6 is a pathogenic bacterium with a genome sequence length of 3.28 Mb [36]. Previous studies have shown that this genome has three anomaly gene clusters, VVGI-1 (2,438,377-2,605,507bp), VVGI-2 (355,728-395,914bp) and VVGI-3 (3,248,897-3,281,945bp) [37].

With the combination of (50K, 6-mer), we predicted four genomic regions that have the probability values less than -5500, i.e., 0.30-0.40, 0.75-0.80, 2.45-2.60 and 3.23-3.28 (Figure 2(d)). The region of 0.30-0.40 contains 93 genes, most of them are hypothetical proteins, but it also contains type IV secretory proteins, and phage integrase. The region of 2.45-2.60 contains 131 genes, where it contains nine transposase, integrase and prophage antirepressor.

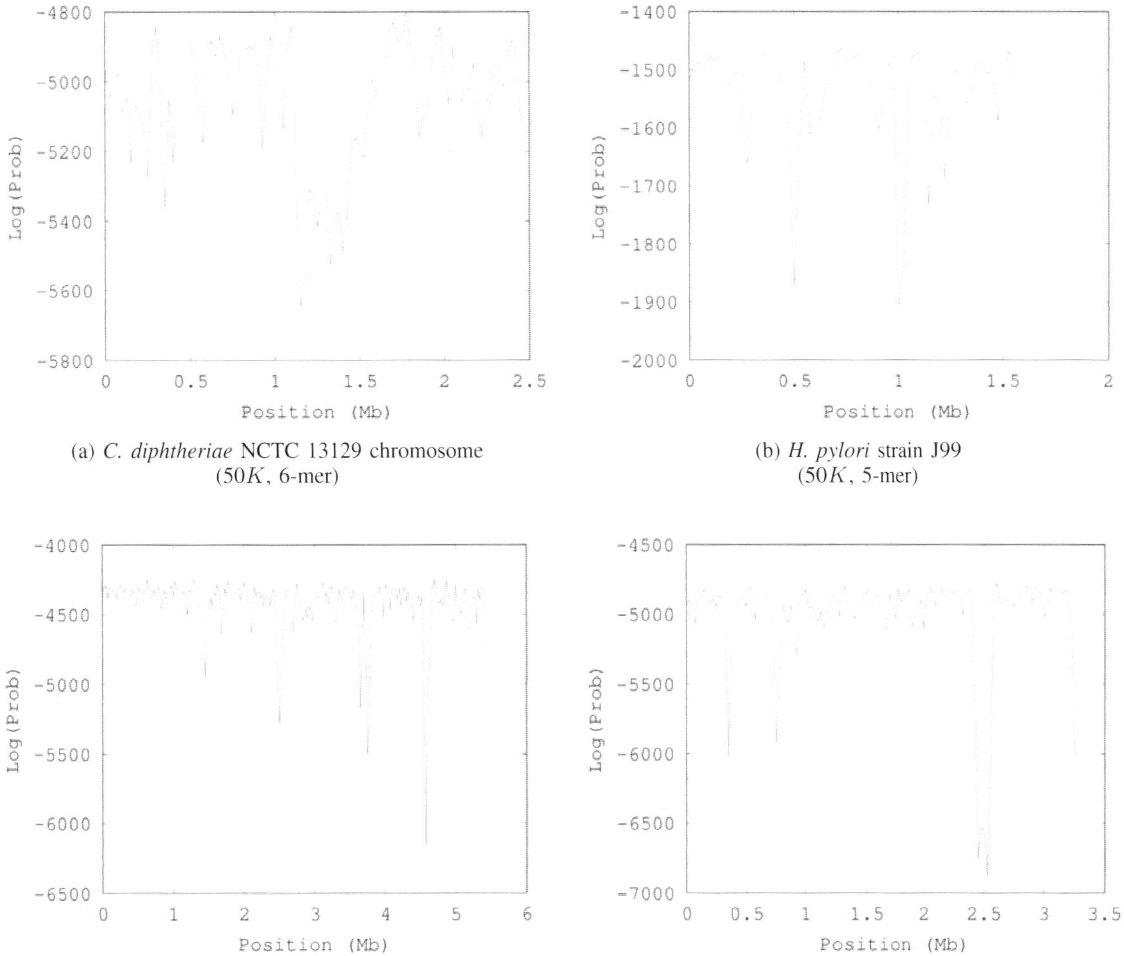

(a) *C. diphtheriae* NCTC 13129 chromosome (50K, 6-mer)

(b) *H. pylori* strain J99 (50K, 5-mer)

(c) *R. palustris* CGA009 (50K, 6-mer)

(d) *V. vulnificus* CMCP6 chromosome I (50K, 6-mer)

Fig. 2: Graphic representations for the probabilities in the genomes of (A) *C. diphtheriae* NCTC 13129 chromosome; (B)*H. pylori* strain J99; (C) *R. palustris* CGA009 and (D) *V. vulnificus* CMCP6 chromosome I

4. Conclusions and Discussion

In this paper, we have implemented our genome anomaly detection algorithm for detecting alien gene clusters in bacterial and archaeal genomes. The algorithm is parameterized by the block size B and k-mer word. Our experiments on five genome sequences have shown that the block size of 50 K, and 6-mer word (5-mer for small genome size) have better prediction power than smaller block sizes and smaller k-mer word. Most of our detected anomalous gene clusters are consistent with previous studies, which showed horizontal gene transfer evidence through case by case studies.

While the genome sequence based detection algorithm can be applied to any microbial genome, it remains to be a challenging task for the following scenarios: (1) The genome sequences of the donor and recipient species could be similar, thus, making it difficult to detect alien gene clusters in the core genome. (2) The alien genome sequence in the host genome can be ameliorated, a process that makes the sequence composition (or codon usage) of the alien genomic region be similar to that of the core genome. A recent large scale genomic study of bacterial and archaeal genomes have supported the existence of amelioration. Again, the anomaly detection algorithm is incapable of detecting such ameliorated regions. (3) Not all biased genome sequences are horizontally transferred. Instead, they could be essential to host genomes and happen to be biased. Highly expressed genes, such as ribosomal related genes, chaperonin genes, transcription and termination factor genes, energy

Table 2: Alien gene clusters detected by our algorithm with supporting information

Parameter	Cut-off	Predicted Region (Mb)	Genes Covered	Reported Region (Mb)	Function
C. glutamicum ATCC 13032					
(50k, 6 mer)	-6000	1.75-2.00	236	1.78-1.99 [37]	Mainly hypothetical proteins, with one integrases, one phage genes, seven tRNAs, and two transposases
C. diphtheriae NCTC 13129 chromosome					
(50k, 6 mer)	-5300	0.15-0.20	57	0.15-0.19 [28]	Corynephage and diphtheria toxin gene
		0.25-0.30	48	0.25-0.26 [28]	Putative iron transport system
		0.35-0.40	53	-	Transposase, insertion element
		1.12-1.52	351	-	Contains nine tRNA genes, and four putative transposase genes.
H. pylori strain J99					
(50k, 5 mer)	-1800	0.50-0.55	49	0.50-0.54 [34]	cag island proteins
		1.00 - 1.10	83	-	Majority of putative proteins with several integrases
R. palustris CGA009					
(50k, 6 mer)	-5000	2.40-2.60	160	2.48-2.57 [35]	Type IV secretion system subunit and conjugal transfer proteins
		3.70-3.80	78	3.73-3.80 [35]	Hypothetical proteins
		4.55-4.70	177	4.58-4.68 [35]	Conjugal transfer proteins, multidrug efflux transporters, integrases and tRNAs
V. vulnificus CMCP6 chromosome I					
(50k, 6 mer)	-5500	0.30-0.40	93	0.35-0.40 [37]	Hypothetical proteins, type IV secretory protein, phage integrase
		0.75-0.80	55	-	Ribosomal proteins
		2.45-2.60	131	2.44-2.60 [37]	Hypothetical proteins, transposase, integrase and phage
		3.23-3.28	46	3.25-3.28 [37]	Transporter protein, transposase, phage and hypothetical proteins

metabolism genes, recombination and repair genes, and electron transport genes, have the characteristics of sequence bias. In this scenario, the detected anomalous regions are not horizontally transferred. Given that, additional information might be incorporated into the algorithm to handle these scenarios.

Our experiments have shown that large block size (e.g., $50k$) and 6-mer (or 5-mer) word have fairly good detection power. Such settings could be adjusted slightly for detecting any other genomes, depending on the genome sequence size. The threshold values for selecting alien gene clusters in the genomes is hard to determine since different genomes have various alien genome percentage. It has been reported that some genomes could have up to half of genome sequences transferred from other organisms. One of possible solution to that is to use supervised learning approach to obtain threshold value for new genomes.

References

[1] V. Hodge and J. Austin, "A Survey of Outlier Detection Methodologies," *Artificial Intelligence Review*, vol. 22, pp. 85-126, 2003.

[2] I. Chen and D. Dubnau, "DNA uptake during bacterial transformation," *Nature Reviews Microbiology*, vol. 2, no. 3, pp. 241-249, 2004.

[3] J. Lederberg and E. L. Tatum, "Gene recombination in Escherichia coli," *Nature*, vol. 158, no. 4016, article 558, 1946.

[4] C. Canchaya, G. Fournous, S. Chibani-Chennoufi, M.-L. Dillmann, and H. Brüssow, "Phage as agents of lateral gene transfer," *Current Opinion in Microbiology*, vol. 6, no. 4, pp. 417-424, 2003.

[5] A. S. Lang, O. Zhaxybayeva, and J. T. Beatty, "Gene transfer agents: phage-like elements of genetic exchange," *Nature Reviews Microbiology*, vol. 10, no. 7, pp. 472-482, 2012.

[6] D. Che, Hasan MS, Chen B, "Identifying pathogenicity islands in bacterial pathogenomics using computational approaches," *Pathogens* 3:36-56, 2014

[7] Lowe, T.M.; Eddy, S.R. "tRNAscan-SE: A program for improved detection of transfer RNA genes in genomic sequence." *Nucleic Acids Res.* 25, 955-964, 1997

[8] P. Wan, D. Che, "A Computational Framework for Tracing the Origins of Genomic Islands in Prokaryotes," *International Scholarly Research Notices, Hindawi Publishing Corporation,*, vol. 2014(2014), article 732857, Oct 2014

[9] Winstanley C, Langille MG, et al. "Newly introduced genomic prophage islands are critical determinants of in vivo competitiveness in the liverpool epidemic strain of Pseudomonas aeruginosa." *Genome Res.* 19:12-23, 2009.

[10] Hacker, J.; Bender, L.; Ott, M.; Wingender, J.; Lund, B.; Marre, R.; Goebel, W. "Deletions of chromosomal regions coding for fimbriae and hemolysins occur in vitro and in vivo in various extraintestinal Escherichia coli isolates," *Microb. Pathog.* 8, 213-225, 1990.

[11] Brinkman FS, Hancock RE, Stover CK, "Sequencing solution: use volunteer annotators organized via internet." *Nature.* 406:933, 2000.

[12] Winsor GL, Khaira B, Van Rossum T, Lo R, Whiteside MD, Brinkman FS. "The Burkholderia genome database: facilitating flexible queries and comparative analyses," *Bioinformatics.* 24:2803-2804, 2008.

[13] Klockgether J, Munder A, Neugebauer J, Davenport CF, Stanke F, Larbig KD, Heeb S, Schock U, Pohl TM, Wiehlmann L, et al. "Genome diversity of Pseudomonas aeruginosa PAO1 laboratory strains," *J. Bacteriol.* 192:1113-1121, 2010.

[14] Kurtz S, Phillippy A, Delcher AL, Smoot M, Shumway M, Antonescu C, Salzberg SL. "Versatile and open software for comparing large genomes," *Genome Biol.* 5:R12, 2004.

[15] G. S. Vernikos and J. Parkhill, "Interpolated variable order motifs for identification of horizontally acquired DNA: revisiting the Salmonella pathogenicity islands," *Bioinformatics,* vol. 22, no. 18, pp. 2196-2203, 2006.

[16] I. Rajan, S. Aravamuthan, and S. S. Mande, "Identification of compositionally distinct regions in genomes using the centroid method," *Bioinformatics,* vol. 23, no. 20, pp. 2672-2677, 2007.

[17] D. Che, C. Hockenbury, R. Marmelstein, and K. Rasheed, "Classification of genomic islands using decision trees and their ensemble algorithms," *BMC Genomics,* vol. 11, supplement 2, article S1, 2010.

[18] H. Wang, J. Fazekas, M. Booth, Q. Liu, and D. Che, "An integrative approach for genomic island prediction in prokaryotic genomes," in *Bioinformatics Research and Applications,* J. Chen, J. Wang, and A. Zelikovsky, Eds., vol. 6674, pp. 404-415, Springer, Berlin, Germany, 2011.

[19] Q. Tu and D. Ding, "Detecting pathogenicity islands and anomalous gene clusters by iterative discriminant analysis," *FEMS Microbiology Letters,* vol. 221, no. 2, pp. 269-275, 2003.

[20] R. Merkl, "SIGI: score-based identification of genomic islands," *BMC Bioinformatics,* vol. 5, article 22, 2004.

[21] M. G. I. Langille, W. W. L. Hsiao, and F. S. L. Brinkman, "Evaluation of genomic island predictors using a comparative genomics approach," *BMC Bioinformatics,* vol. 9, article 329, 2008.

[22] H.-Y. Ou, X. He, E. M. Harrison et al., "MobilomeFINDER: web-based tools for in silico and experimental discovery of bacterial genomic islands," *Nucleic Acids Research,* vol. 35, no. 2, pp. W97-W104, 2007.

[23] D. Che, M. S. Hasan, H. Wang, J. Fazekas, J. Huang, and Q. Liu, "EGID: an ensemble algorithm for improved genomic island detection in genomic sequences," *Bioinformation,* vol. 7, no. 6, pp. 311-314, 2011.

[24] Shrivastava, S.; Reddy Ch, V.; Mande, S.S. INDeGenIUS, "A new method for high-throughput identification of specialized functional islands in completely sequenced organisms," *J. Biosci.* 35, 351-364, 2010.

[25] Hsiao, W.; Wan, I.; Jones, S.J.; Brinkman, F.S. "IslandPath: Aiding detection of genomic islands in prokaryotes," *Bioinformatics.* 19, 418-420, 2003.

[26] D. Che, H. Wang, J. Fazekas, B. Chen."An Accurate Genomic Island Prediction Method for Sequenced Bacterial and Archaeal Genomes,", *J Proteomics Bioinform,* 7:8, 2014.

[27] Ikeda M, Nakagawa S. "The Corynebacterium glutamicum genome: features and impacts on biotechnological processes. Applied Microbiology and Biotechnology," 62(2-3):99-109. doi: 10.1007/s00253-003-1328-1. pmid:12743753, 2003.

[28] Cerdeño-Tàrraga AM, Efstratiou A, Dover LG, Holden MT, Pallen M, et al. "The complete genome sequence and analysis of Corynebacterium diphtheriae NCTC13129," *Nucleic Acids Res* 31: 6516-6523, 2003.

[29] Alm RA, Ling LS, Moir DT, King BL, Brown ED, Doig PC, Smith DR, Noonan B, Guild BC, deJonge BL, et al. "Genomic-sequence comparison of two unrelated isolates of the human gastric pathogen Helicobacter pylori," *Nature.* 397:176-80, 1999.

[30] Larimer FW, Chain P, Hauser L, Lamerdin J, Malfatti S, Do L, et al. "Complete genome sequence of the metabolically versatile photosynthetic bacterium Rhodopseudomonas palustris," *Nature biotechnology.* 22(1):55âĂŞ61. doi: 10.1038/nbt923. pmid:14704707, 2004.

[31] Chen CY, Wu KM, Chang YC, Chang CH, Tsai HC, Liao TL, et al. "Comparative genome analysis of Vibrio vulnificus, a marine pathogen," *Genome research,* 13(12):2577âĂŞ2587. doi: 10.1101/gr.1295503. pmid:14656965, 2003.

[32] Kalinowski J, Bathe B, Bartels D, Bischoff N, et al. "The complete Corynebacterium glutamicum ATCC 13032 genome sequence and its impact on the production of L-aspartate-derived amino acids and vitamins," *J Biotechnol.* 104: 5-25, 2003.

[33] M. G. I. Langille and F. S. L. Brinkman, "IslandViewer: an integrated interface for computational identification and visualization of genomic islands," emphBioinformatics, vol. 25, no. 5, pp. 664-665, 2009.

[34] Covacci, A., Telford, J.L., Del Giudice, G., Parsonnet, J. & Rappuoli, R, "Helicobacter pylori virulence and genetic geography," *Science* 284, 1328-1333, 1999.

[35] Zhang C, Zhang R, "Genomic islands in Rhodopseudomonas palustris," *Nat Biotechnol* 22: 1078-1079, 2004

[36] Kim, Y.R., Lee, S.E., Kim, C.M., Kim, S.Y., Shin, E.K., Shin, D.H., et al. "Characterization and pathogenic significance of Vibrio vulnificus antigens preferentially expressed in septicemic patients," *Infect Immun* 71: 5461-5471, 2003.

[37] Zhang, R. & Zhang, C. T. "A systematic method to identify genomic islands and its applications in analyzing the genomes of Corynebacterium glutamicum and Vibrio vulnificus CMCP6 chromosome I." *Bioinformatics* 20, 612-622, 2004.

PGAR: ASD Candidate Gene Prioritization System Using Expression Patterns

Steven Cogill and Liangjiang Wang
Department of Genetics and Biochemistry, Clemson University, Clemson, SC, USA

Abstract - *PGAR is an autism spectrum disorder (ASD) candidate gene prioritization web-based tool. It is built on a database which houses machine learning and co-expression analysis for a majority of known human genes. Users submit a gene list, and for each gene, a classification score from the machine learning model is retrieved. A prioritized gene list is returned based on the classification score with links to gene profiles with co-expression analysis. The system is novel in its use of expression patterns, which allows for prioritization of non-coding RNA genes and genes lacking functional annotation. The user-friendly design, high accuracy classification model, and depth of information for all genes make PGAR useable and invaluable for researchers studying ASD.*

Keywords: autism, prioritization, machine learning, co-expression, web-based tool

1 Introduction

Autism spectrum disorder (ASD) is a heterogeneous group of disorders convergent on a behavioral phenotype with a strong genetic component [1]. Hundreds of genes are currently associated with ASD, and given an etiology that is still not definitively known, many more genes are hypothesized to be associated [2]. Current high-throughput screening methods such as genome-wide association studies produce multiple targets, which are not feasible to research on an individual level. Therefore, prioritization is needed. However, many candidate gene prioritization systems have focused on annotated protein-coding genes through the use of protein-protein interaction networks and literature mining [3,4]. This neglects the other gene types. This is very prominent in ASD for one gene type in particular, long non-coding RNA genes defined as genes, which have transcripts greater than 200 nucleotides that do not code for protein. These genes have been shown to be both developmental regulators, highly expressed in the brain, and differentially expressed in ASD cases [5,6]. To our knowledge, our system is unique in its use of expression data from developmental brain tissue for ASD gene prioritization. Here we present the Prioritization System of Genes for Autism Risk (PGAR); a prioritization system which employs support vector machines (SVM) and co-expression network analysis to gene expression profiles to prioritize and annotate gene lists.

PGAR is a web tool with a database backend which facilitates ASD research through the identification of high priority targets within human gene lists. To begin the analysis users submit a gene list. They have the option of pasting an existing list into the field under the "Paste Gene List:" heading or they can submit a text file (Figure 1). Currently PGAR supports gene identifiers in the formats of ENSEMBL IDs [7] and gene symbols as dictated by the HUGO Gene Nomenclature Committee [8]. Alternatively, users can provide loci. For instance, if a region from a copy number variant study is identified, users can simply indicate that region in the input field, and PGAR will return a prioritized list of all the available genes in that region. The format is as follows:

chromosome:start position to end position

For example, 1:10,000-50,000 would refer to all genes on chromosome 1 which overlap or are between positions 10,000 and 50,000. When the loci list is submitted, the user is given the option of either prioritizing the loci as a group or as individual lists.

2 Database

Our database uses a star schema with the gene profile table at the center. Each gene was assigned a unique ID for the PGAR system, and the location and the type of the gene were documented in this center table. For this current iteration of the system, these annotations were from GENCODE [9]. This schema allows for an extensible database in that there are three distinct table groups linked to the main table. One section is comprised of lookup tables for potential identifiers such as the ENSEMBL ID. The second group comprises machine learning results, and the third group is made up of co-expression data. This architecture allows for multiple analyses and identifications of the genes in our system. Currently there is one machine learning analysis and co-expression analysis respectively in our system. The two analyses were run using the same expression dataset and list of known ASD risk genes. These analyses were from previous studies using the BrainSpan dataset, which consists of 524 samples from 8 weeks post conception to 40 years of age and 26 brain structures [10]. The known ASD risk gene list used in the studies was compiled from three different resources [2,11,12]. The dataset has values for all the genes in the GENCODE v10 build, and genes not in the most recent build (v24) were removed leaving 46,782 genes currently in PGAR. Therefore, we have prioritization information for the majority

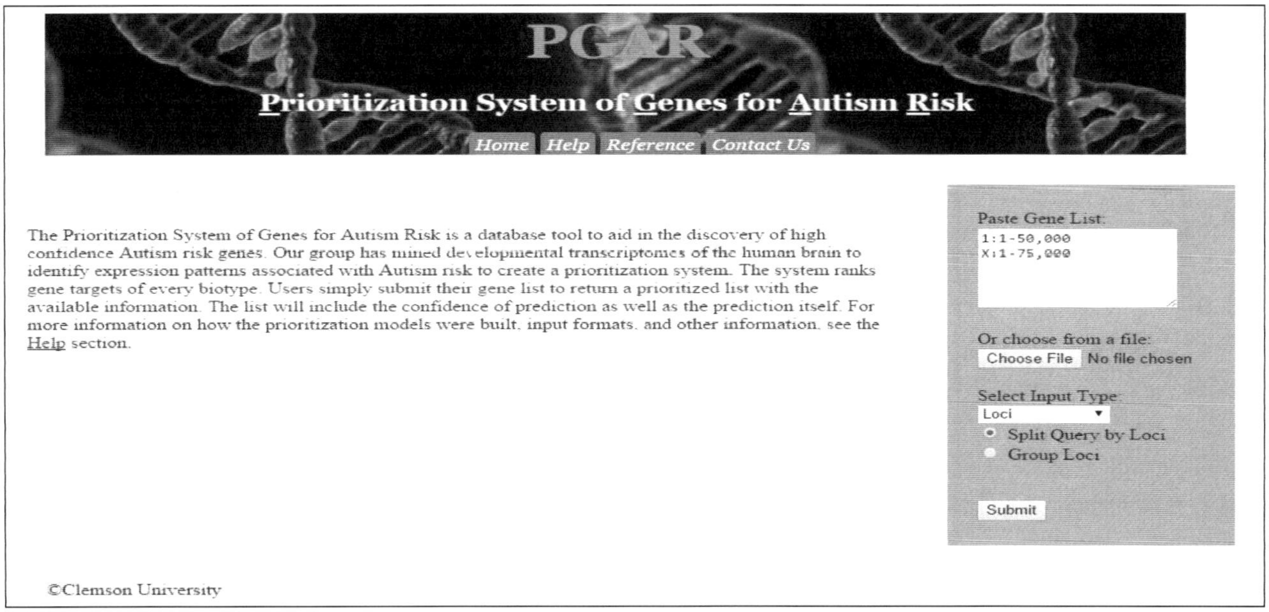

Figure 1: Screenshot of the PGAR home page. A brief introduction of the system is provided. The panel on the right provides options for the uploading of gene lists for prioritization.

of known genes of all types in the human genome based on their expression patterns in the developing human brain.

3 Supervised machine learning model

The values used for prioritization are from a previous study (under review) performed by this group. In this study, we developed a supervised machine learning model. Briefly, the BrainSpan dataset was used, and the 524 features in the dataset were reduced down to 15 features using a wrapper method with the SVM algorithm and a best-first search method. Once the model was generated, all of the available genes in the dataset were put through the model to generate an SVM output value. This value is what the genes are prioritized on. The output itself has a sign value associated with it, and this determines the classification of the gene as either ASD risk or non-ASD risk which is what is shown in the output. To assign a meaningful numeric value, a confidence score is given for the classification. This is on a scale of 0-1 for both negative and positive classifications. For example, a gene with high confidence for ASD risk could have a value of 0.9, and a gene with high confidence for non-ASD risk could also have a value of 0.9. These values are based on the range of outputs for the genes used to train the model.

4 Co-expression network

In another study we sought to provide functional annotation of lncRNAs through weighted gene co-expression analysis (WGCNA) (in manuscript). In that study, we curated the BrainSpan dataset down to 20,456 genes with the highest covariance sums, and then clustered those genes based on co-expression into modules. Further enrichment analyses was performed. It then became our goal to find a way to incorporate this information into the PGAR system. Therefore using the adjacency matrix generated in the study, we summed the weighted connectivity between each gene in the curated dataset and all the ASD seed genes. This provided a means of measuring co-expression with known ASD genes and offered further insight into the role of the gene in ASD. Next we analyzed each module as a means of providing partial functional annotation for the genes within our system through their module assignment. We calculated the enrichment of ASD genes within the module using a Fisher's exact test for the frequencies of ASD genes within the module and those for the total set. Then we performed term enrichment analysis on the modules using the DAVID bioinformatics tool [13].

5 PGAR output

After submission of a gene list, the user is directed to a page containing a results table of the prioritized gene list (Figure 2). The far left column is the PGAR ID, which is the system's unique identifier. The next column is the gene name, which is the identifier in the submitted list, and the third column is the list rank. This number is the prioritized rank within the submitted gene list, and it should be noted that this is not the rank within the entire system. The next column is the classification as an ASD risk or non-ASD risk gene, and it should be noted that this is the candidate classification prediction by the PGAR system. Finally, the confidence of

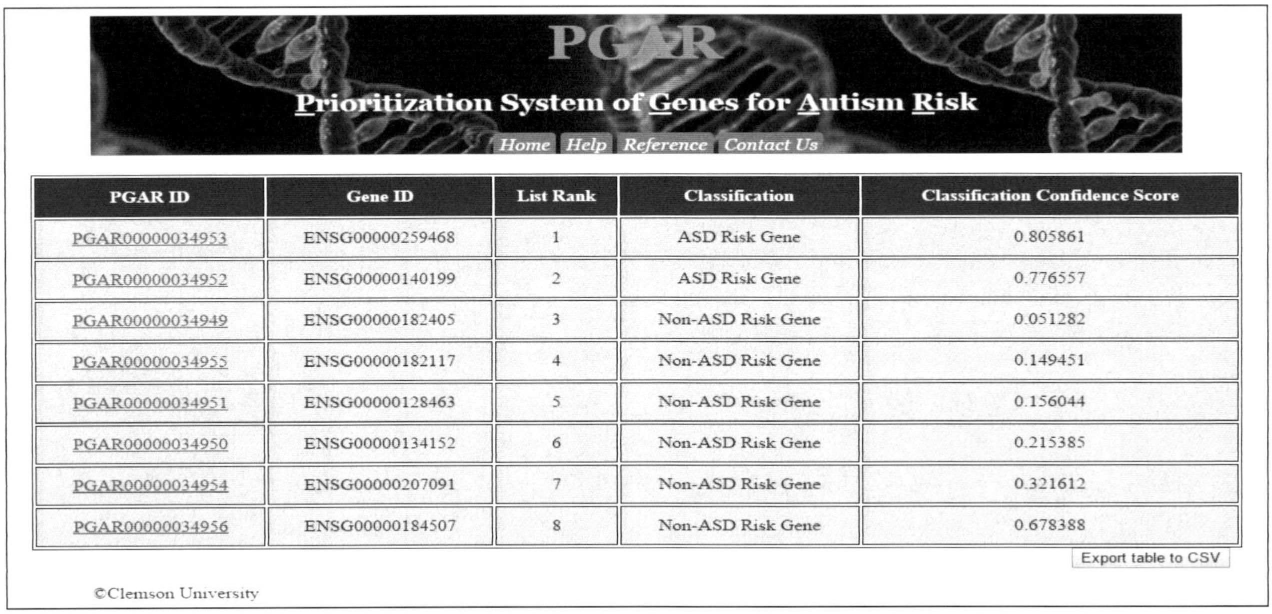

PGAR ID	Gene ID	List Rank	Classification	Classification Confidence Score
PGAR00000034953	ENSG00000259468	1	ASD Risk Gene	0.805861
PGAR00000034952	ENSG00000140199	2	ASD Risk Gene	0.776557
PGAR00000034949	ENSG00000182405	3	Non-ASD Risk Gene	0.051282
PGAR00000034955	ENSG00000182117	4	Non-ASD Risk Gene	0.149451
PGAR00000034951	ENSG00000128463	5	Non-ASD Risk Gene	0.156044
PGAR00000034950	ENSG00000134152	6	Non-ASD Risk Gene	0.215385
PGAR00000034954	ENSG00000207091	7	Non-ASD Risk Gene	0.321612
PGAR00000034956	ENSG00000184507	8	Non-ASD Risk Gene	0.678388

Figure 2: Screenshot of the PGAR results page. Below the banner, a table is generated for the prioritized gene list with links to the system profiles for each gene.

this classification is given in the last column. For loci submissions, the locus searched is given at the top of the table and for loci prioritized separately, a table is generated for each locus. For all tables, there is a button to export the table to a ".csv" file located at the bottom right of the table.

Each PGAR ID is a link to the profile for that gene. In that profile, general information for the gene is provided which includes the gene symbol, location, and type (Figure 3). Next the classification and confidence from the machine learning based prioritization is shown, which for the current prototype is from the SVM model outlined above. Following that, the co-expression information is displayed. This consists of the gene's weighted connectivity with known ASD genes as well as the gene's module assignment. It should be noted that not all genes have associated co-expression information because as stated previously, to form the co-expression network, the original dataset was curated. However given the nature of our machine learning approach, this does not necessarily preclude them from being high priority candidates or there utility as negative instances. Therefore, genes without co-expression data were not removed from the system.

Each module assignment on the profile pages links to a profile for that module. That profile includes basic information about the module which is the number of genes within the module and ASD gene enrichment *P*-value (Figure 4). The profile also includes a table of the enrichment terms for the module. At the top of the list is a link to the source for the enrichment terms which currently is the DAVID bioinformatics server. For each listing, the term itself, its broader category designation, fold enrichment, and *P*-value are given.

6 Software validation

Future plans for the PGAR software include multiple machine learning methods as options or considered together in an ensemble approach, but currently the system uses the high performance model built using the previously described supervised SVM approach. The model boasts a 77% classification accuracy for ASD vs non-ASD risk genes and has been demonstrated to prioritize known ASD genes highly. Given that a majority of known ASD risk genes were used in our machine learning and co-expression studies, testing of the prioritization system requires new data. We are currently in collaboration with another group at the Greenwood Genetic Center, which is independently testing the PGAR system using copy number variant studies where a larger region was associated with ASD and that region was subsequently parsed down through a process of elimination. Given that the novel risk gene(s) is known, the performance of the system can be evaluated by its ability to classify the gene(s) as an ASD risk gene and to prioritize it highly in the gene list for the region. The software is currently residing on a test server at http://scogill.people.clemson.edu/PGAR.php. We are in the process of testing all of the utilities as well as compatibility with various browsers.

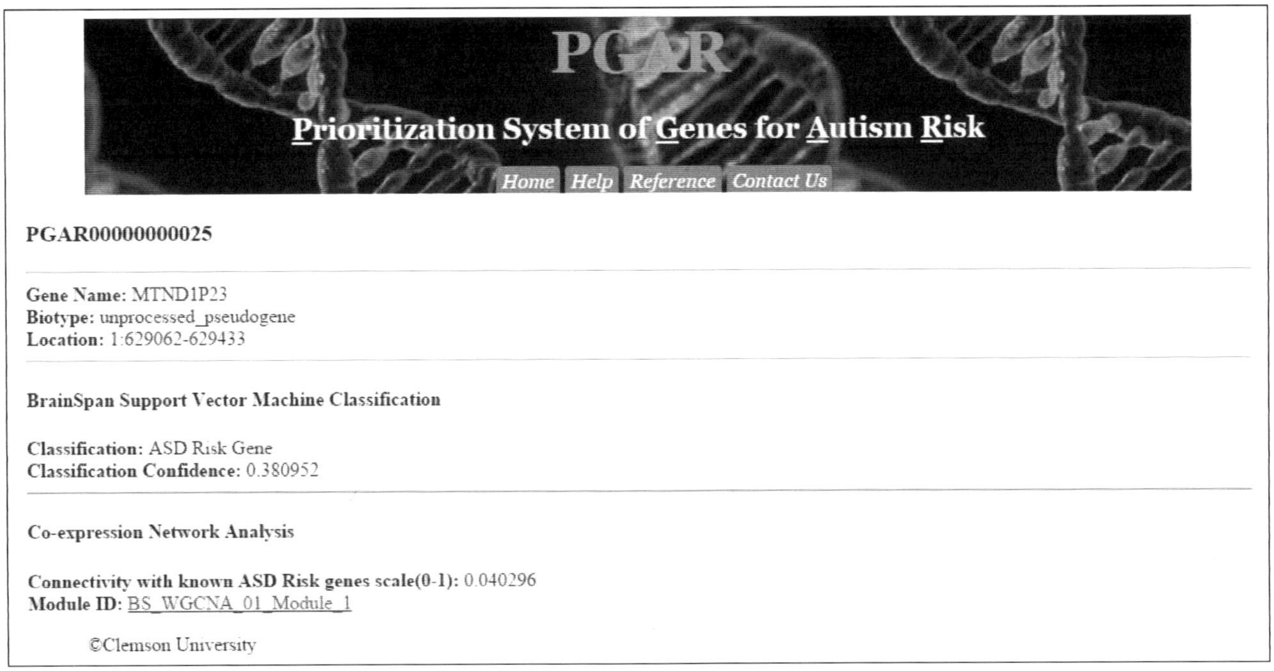
PGAR
Prioritization System of Genes for Autism Risk
Home Help Reference Contact Us

PGAR00000000025

Gene Name: MTND1P23
Biotype: unprocessed_pseudogene
Location: 1:629062-629433

BrainSpan Support Vector Machine Classification

Classification: ASD Risk Gene
Classification Confidence: 0.380952

Co-expression Network Analysis

Connectivity with known ASD Risk genes scale(0-1): 0.040296
Module ID: BS_WGCNA_01_Module_1

©Clemson University

Figure 3: Screenshot for the PGAR gene profile page. This page shows the profile for an unprocessed pseudogene, which is in module 1 for the co-expression analysis.

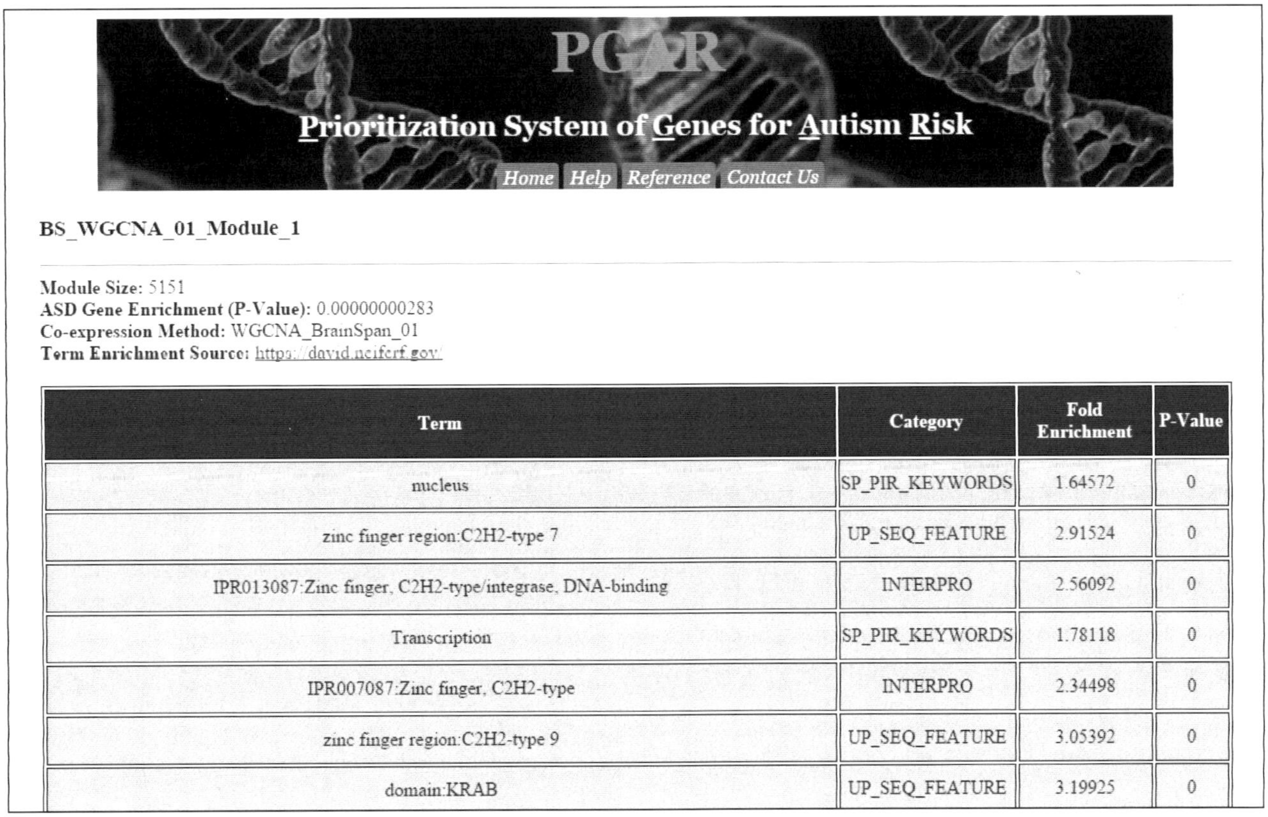
PGAR
Prioritization System of Genes for Autism Risk
Home Help Reference Contact Us

BS_WGCNA_01_Module_1

Module Size: 5151
ASD Gene Enrichment (P-Value): 0.00000000283
Co-expression Method: WGCNA_BrainSpan_01
Term Enrichment Source: https://david.ncifcrf.gov/

Term	Category	Fold Enrichment	P-Value
nucleus	SP_PIR_KEYWORDS	1.64572	0
zinc finger region:C2H2-type 7	UP_SEQ_FEATURE	2.91524	0
IPR013087:Zinc finger, C2H2-type/integrase, DNA-binding	INTERPRO	2.56092	0
Transcription	SP_PIR_KEYWORDS	1.78118	0
IPR007087:Zinc finger, C2H2-type	INTERPRO	2.34498	0
zinc finger region:C2H2-type 9	UP_SEQ_FEATURE	3.05392	0
domain:KRAB	UP_SEQ_FEATURE	3.19925	0

Figure 4: Screenshot of PGAR module summary page. The page gives a brief summary of the module attributes and displays a table of the enrichment terms associated with the module.

7 Other prioritization systems

ASD is currently a high profile area of study. There are many data repositories of ASD risk genes including AutKB [11] and SFARI [12], which characterize genes based on empirical evidence from previous studies. While this is useful, it does not allow for the identification of novel candidate genes. Many of the existing popular prioritization systems such as DADA [3], ENDEAVOR [14], and GeneMANIA [15] use broad expression networks, existing annotations, or rely upon protein-protein interactions. Our system is novel in that it uses brain developmental data in a targeted approach to ASD gene prioritization. ASD is a neurodevelopmental disorder, and the study of expression patterns in developing brain tissue is essential to identifying high priority candidate genes. This specialization for ASD gives our system a performance advantage over existing systems, and we believe this will lead to the identification of many novel non-coding ASD candidate genes.

8 Conclusions

This system offers a novel approach to the identification of high-confidence ASD candidate genes. The system is easy to use with a simple interface in that input is in the form of gene lists. In addition to the software introduction given here, our site also provides a 'Help' section found in the banner at the top. One of the future goals, is the expansion of the allowable inputs to allow the user more options for data analysis. This may include transcript IDs as well as microarray probes. We also plan to expand the system in several areas. As more known ASD risk genes are discovered or genes within are current set are curated, we will update and rerun our analyses to determine any significant performance benefits in the form of higher classification accuracy or identification of more relevant interactions within the co-expression network. We also plan to increase the number of expression datasets used. The BrainSpan dataset is unique in its comprehensive coverage of the developmental brain transcriptome, but we have begun looking for other suitable expression datasets to be incorporated into our prioritization analysis. The use of new expression datasets may allow us to increase the number of genes within our system. The database architecture allows for multiple machine learning models and co-expression networks. With additional expression datasets, we can add more analysis data to the system and potentially improve on the existing entries. We are currently in the process of testing the system, and in the future we will start updating PGAR.

9 References

[1] American Psychiatric Association. "Diagnostic and statistical manual of mental disorders" (5th ed.). Washington, DC: Author, 2013.

[2] B. Abrahams, D. Arking, et al., "SFARI Gene 2.0: a community-driven knowledgebase for the autism spectrum disorders (ASDs)", *Molecular Autism*, vol. 4, no. 1, p. 36, 2013.

[3] S. Erten, et al., "DADA: Degree-Aware Algorithms for Network-Based Disease Gene Prioritization", *BioData Mining*, vol. 4, no. 1, p. 19, 2011.

[4] D. Hristovski, et al., "Using literature-based discovery to identify disease candidate genes", *International Journal of Medical Informatics*, vol. 74, no. 2-4, pp. 289-298, 2005.

[5] T. Derrien, et al., "The GENCODE v7 catalog of human long noncoding RNAs: Analysis of their gene structure, evolution, and expression", *Genome Research*, vol. 22, no. 9, pp. 1775-1789, 2012.

[6] M. Ziats and O. Rennert, "Aberrant Expression of Long Noncoding RNAs in Autistic Brain", *Journal of Molecular Neuroscience*, vol. 49, no. 3, pp. 589-593, 2012.

[7] P. Flicek, et al., "Ensembl 2014." *Nucleic Acids Research*, 42 no. 1, pp. D749-D755, 2014

[8] K. Gray, et al., "Genenames.org: the HGNC resources in 2013", *Nucleic Acids Research*, vol. 41, no. 1, pp. D545-D552, 2012.

[9] J. Harrow, et al., "GENCODE: The reference human genome annotation for The ENCODE Project", *Genome Research*, vol. 22, no. 9, pp. 1760-1774, 2012.

[10] M. Hawrylycz, et al., "An anatomically comprehensive atlas of the adult human brain transcriptome", *Nature*, vol. 489, no. 7416, pp. 391-399, 2012.

[11] S. Basu, et al., "AutDB: a gene reference resource for autism research", *Nucleic Acids Research*, vol. 37, pp. D832-D836, 2009.

[12] L. Xu, et al., "AutismKB: an evidence-based knowledgebase of autism genetics", *Nucleic Acids Research*, vol. 40, no. 1, pp. D1016-D1022, 2011.

[13] D. Huang, et al., "Systematic and integrative analysis of large gene lists using DAVID bioinformatics resources", *Nat Protoc*, vol. 4, no. 1, pp. 44-57, 2008.

[14] L. Tranchevent, et al., "ENDEAVOUR update: a web resource for gene prioritization in multiple species", *Nucleic Acids Research*, vol. 36, pp. W377-W384, 2008.

[15] D. Warde-Farley, et al., "The GeneMANIA prediction server: biological network integration for gene prioritization and predicting gene function", *Nucleic Acids Research*, vol. 38, pp. W214-W220, 2010.

Information-theoretic Interestingness Measures for Cross-Ontology Data Mining

Prashanti Manda[1*], and Fiona McCarthy[2], and Bindu Nanduri[3], and Hui Wang[3], and Susan M. Bridges[4]
[1]Department of Biology, University of North Carolina at Chapel Hill, Chapel Hill, NC, USA
[2]Department of Veterinary Science and Microbiology, University of Arizona, Tucson, AZ, USA
[3]Department of Basic Sciences, College of Veterinary Medicine, Mississippi State University, MS, USA
[4]Information Technology and Systems Center, University of Alabama, Huntsville, AL, USA
*Corresponding author

Abstract— *Community annotation of biological entities with concepts from multiple bio-ontologies has created large and growing repositories of ontology-based annotation data with embedded implicit relationships among orthogonal ontologies. Development of efficient data mining methods and metrics to mine and assess the quality of the mined relationships has not kept pace with the growth of annotation data. In this study, we present a data mining method that uses ontology-guided generalization to discover relationships across ontologies along with a new interestingness metric based on information theory. We apply our data mining algorithm and interestingness measures to datasets from the Gene Expression Database at the Mouse Genome Informatics as a preliminary proof of concept to mine relationships between developmental stages in the mouse anatomy ontology and Gene Ontology concepts (biological process, molecular function and cellular component). In addition, we present a comparison of our interestingness metric to four existing metrics. Ontology-based annotation datasets provide a valuable resource for discovery of relationships across ontologies. The use of efficient data mining methods and appropriate interestingness metrics enables the identification of high quality relationships.*

Keywords: gene ontology, annotations, association rule mining, interestingness measures, cross-ontology

1. Introduction

The wide spread use of ontologies to describe data has led to the availability of large ontology-based datasets where different ontologies are often used to describe distinct characteristics of entities. For example, in the biological and bio-medical domain, the Gene Ontology might be used to describe the biological processes of a gene product while an anatomy ontology is used to specify the location of expression. The integration of these distinct ontology-based datasets lends itself to the discovery of interesting relationships between the ontologies (cross-ontology relationships). These relationships enable data and information integration and lead to the discovery of patterns not evident from individual datasets. For example, cross-ontology relationships mined from gene expression and annotation data can be used to answer "big picture" questions such as "What biological processes are typically expressed in the mouse brain?"

The abundance of ontology-based annotations is accompanied by a dearth of efficient data mining techniques that can discover biologically relevant relationships from the data. One of the drawbacks of data mining techniques such as association rule mining is the retrieval of large number of relationships that need to be prioritized or ranked based on domain knowledge. Existing ranking metrics are either unsuitable for ontology based relationships [1] or do not accommodate domain knowledge. The gap in techniques to mine and rank biological ontology-based relationships forms the motivation for this work.

This paper focuses on data mining methods for the integration and mining of ontology-based annotation datasets and describes a new information theoretic metric to rank the mined cross-ontology relationships (relationships between concepts from different ontologies). Our data mining algorithm integrates ontological datasets and mines cross-ontology association rules to indicate the relationships between two ontologies describing different aspects of biological entities. Note that this form of discovery is distinct from efforts to map concepts across different ontologies describing the same aspects of entities.

An association rule is defined as an implication of the form $x \rightarrow y$ where x (antecedent) and y (consequent) are co-occurring items derived from a transaction set T. In association rules that describe market sales, transactions are sets of items purchased together. In cross-ontology association rules derived from annotation data, each transaction contains a gene name and one or more annotations from each ontology (in this study, GO and anatomy ontologies) and x and y are co-annotated concepts from different ontologies. In the data used for this study, transactions express the involvement of gene products in specific processes/functions/components and expression of the gene product in specific tissues. Additionally, x and y are restricted to single concepts instead of itemsets.

Our data mining algorithm employs subsumption reasoning to mine relationships at multiple levels across the input ontologies. Our previous work on generalization algorithms explored two methods of generalization:

1) Level-by-level generalization [2]. Depth of annotations is used to conduct incremental generalization and mining one level at a time.
2) Generalization to all ancestors via transitive relationships [1]. This generalization method is an improvement over the level-by-level generalization since it does not rely on the depth of annotations as a guide for generalization. Instead, generalization is conducted in a single step where all annotations are supplemented with all their ancestors in the ontology. The mining step is conducted only once after the generalization process to improve efficiency.

These algorithms have been applied to GO annotation data and were used to discover relationships across the ontologies of the GO. While our previous methods use the depth of ontology terms to guide generalization, Information content has been shown to be a more accurate indicator of ontology term specificity as compared to depth in the ontology [3], [4]. This is because ontologies evolve over time and different sections of an ontology are developed to different extents depending on the level of available scientific knowledge and the involvement of the specific research community. In the research reported in this paper, we propose a new information theoretic interestingness measure called Integrated Rule Information Content ($IRIC$) to inform ontology-enabled association rule mining from multiple ontologies. $IRIC$ combines the information content of the terms in a rule with the shared information among the terms to accurately assess the interestingness of the rule. $IRIC$ is calculated from the following two components:

1) Normalized Information Content (N_IC): N_IC indicates the information content of ontology terms in a cross-ontology association rule.
2) Normalized Cross-ontology Mutual Information (N_COMI): N_COMI quantifies the information shared by the terms in a cross-ontology association rule.

We apply our data mining algorithm to GO annotation and tissue expression data from the Gene Expression Database at MGI [5] to discover relationships between the GO ontologies and the Mouse Anatomy ontology. N_IC and N_COMI thresholds are used to filter uninformative terms and relationships while $IRIC$ scores are used to rank the remaining relationships.

2. Related Work

Association rule mining has been applied to ontology based mining by several previous studies to discover relationships between one or multiple ontologies [6], [7], [8], [9], [10], [11], [12] . In the majority of these studies, relationships are discovered from a single ontology [6], [7], [8], [9], [11], [12] while some methods can be used for cross-ontology mining as well [10].

While association rule mining in annotation datasets has been explored widely, few studies have focused on developing alternative and appropriate interestingness metrics for ontology based association rules [6], [10], [12]. Faria et al. use association rules to quantify annotation inconsistency by exploring erroneous, incomplete, and inconsistent annotations. Although, Support and Confidence are used as initial interestingness metrics during mining, Faria et al. employ strategies post mining to filter the discovered rules. The metrics (Generic rules, Agreement, Ancestral and descendant distance) use ontology semantics to weed out uninteresting rules. These metrics are similar to the post filtering techniques we have used in our previous work [1].

Another notable work in this area, Benites et al., proposes the idea of comparing the real value of a rule's interestingness with the expected value [10]. Rules with more significant differences in these values are considered more rare and interesting. Paul et al. introduce a suite of metrics based on semantic similarity and ontological distance adapted from existing metrics. These metrics are applied to discover relationships between human phenotypes (HPO terms) and bone dysplasias. While relationships with semantically similar terms are more valuable in Paul et al.'s application, the ontologies we are using for cross-ontology relationships capture different aspects of the objects and need not be semantically similar to be interesting [12].

3. Materials and Methods

This section describes the generalization method and information theoretic interestingness metric we developed for cross-ontology data mining.

3.1 Generalization and mining

As a preprocessing step for the mining algorithm, gene annotations from different ontologies (in this case, anatomy and GO) are combined to build a transaction set. Each transaction in this set contains a gene along with GO and anatomy annotations for the gene. We apply our generalization algorithm to simultaneously generalize terms from all of the ontologies represented in the transaction set [1]. Generalization supplements the annotations in a transaction with all of their ancestors related via transitive relations. The generalized transactions are then processed to remove uninformative terms using an N_IC threshold as described in Section 3.3. The resulting generalized transactions are mined using Christian Borgelt's implementation of the Apriori algorithm [13]. The mined relationships are further filtered using a N_COMI threshold to remove relationships with insufficient shared information. $IRIC$ is then used to rank the remaining relationships.

3.2 Integrated Rule Information Content

Integrated Rule Information Content ($IRIC$) is a novel interestingness measure that combines information content (N_IC) of concepts in a rule with the shared information in the rule (N_COMI). $IRIC$ of a rule $x \rightarrow y$ is defined as

$$IRIC_{x \rightarrow y} = ((\alpha * N\_IC_x) + (\beta * N\_IC_y)) * N\_COMI_{x \rightarrow y}$$

where α , β are weighted coefficients of concepts from the ontologies of x and y, and

$\alpha + \beta = 1$

Concepts from both the ontologies can be weighted equally by setting α and β to 0.5. Alternatively, greater or lower weights can be attributed to concepts from one ontology by modifying α or β. The range of the $IRIC$ measure is [0, 1]. The components used to calculate $IRIC$ are defined below.

3.2.1 Information content of concepts (N_IC)

We define Normalized Information Content (N_IC) of a term t as

$$N\_IC_t = \frac{-logp(t)}{UB(IC)} \text{ where ;}$$

$$p(t) = \frac{|G_t| + \sum_{i=1}^{j} |G_{C_i}|}{|G|} \text{ and ;}$$

G = set of all genes in the transaction set,

G_t = set of genes annotated to t,

$C_i = \{1, 2, \cdots, j\}$ are the descendants of t in the ontology,.

G_{C_i} = set of genes annotated to descendant C_i

Upper bound for IC, $UB(IC) = -log(\frac{1}{|G|})$

Our definition of N_IC is adapted from Shannon's information content [14] to take into account the implicit annotations indicated by subsumption reasoning over the ontology and to make N_IC comparable to other metrics by restricting the range of the metric to [0,1].

a) Cross-ontology Mutual Information: Mutual information of an association rule captures the shared information content and inter-dependence of the antecedent and the consequent in the rule. The mutual information (MI) [15] of an association rule $x \rightarrow y$, where x and y are items from the transaction set, is defined as

$$MI = p(xy) * log_2(\frac{p(xy)}{p(x)p(y)})$$

This definition of MI uses the entire set of transactions as the background to compute the probabilities thus assuming that all transactions contain annotations from all ontologies in the analysis. However many biological datasets incur the problem of missing data where entities are not annotated to all ontologies in the analysis [1]. To address this issue of missing data, we adapted the standard definition of MI to define Normalized Cross-ontology Mutual Information (N_COMI) for assessing the interestingness of cross-ontology multi-level association rules.

We use the following sets in the definition of Normalized Cross-ontology Mutual Information where $x \rightarrow y$ represents a cross-ontology rule with x and y belonging to different ontologies. Note that these sets are subsets of the input transaction set.

1) $X_{(x \rightarrow y)}$ is the set of transactions containing x and at least one term from the ontology of y.
2) $Y_{(x \rightarrow y)}$ is the set of transactions containing y and at least one term from the ontology of x.
3) $COCategory_{x \rightarrow y}$ is the set of transactions containing at least one term from x's ontology and and one from y's ontology.
4) $XY_{x \rightarrow y}$ is the set of transactions which contains both x and y.

The normalized cross-ontology mutual information (N_COMI) of a rule, $x \rightarrow y$ is defined as

$$N\_COMI_{x \rightarrow y} = \frac{p(xy) * log_2 \frac{p(xy)}{p(x)p(y)}}{min((-log_2 p(x)p(x)), -log_2 p(y)p(y)))} \text{ with;}$$

$$p_x = \frac{|X_{x \rightarrow y}|}{|COCategory_{x \rightarrow y}|} ,$$

$$p_y = \frac{|Y_{x \rightarrow y}|}{|COCategory_{x \rightarrow y}|} \text{ and,}$$

$$p_{xy} = \frac{|XY_{x \rightarrow y}|}{|COCategory_{x \rightarrow y}|}$$

3.3 N_IC and N_COMI thresholds

First, uninformative ontology terms are removed from the transaction set after generalization and prior to mining using an N_IC threshold. This step helps avoid mining rules with uninformative terms that occur frequently in the transaction set. These terms are typically closer to the root of the ontology. In our analysis, uninformative terms are terms that are annotated to many genes in the dataset. Examples of uninformative terms in our dataset include organ system and nervous system in the anatomy ontology, *GO:0005623* (cell), *GO:0065007* (biological regulation), and *GO:0005488* (binding) from the Gene Ontology. Selecting an N_IC threshold is a subjective choice and depends on the application of the discovered rules, the ontologies in question, and the annotation dataset.

Second, an N_COMI threshold is selected using Monte Carlo methods. A synthetic dataset containing the same number of transactions as the transaction set is generated using sampling with replacement from the set of all terms in the transaction set. Cross-ontology multi-level rules are mined from the synthetic data and the N_COMI of the rules is calculated. The rules mined from the synthetic data are considered to be False Positives while rules mined from the actual transaction set are 'True Positives'. The False Positives and True Positives are combined and rules are ranked by N_COMI. A N_COMI threshold is selected to yield a desired false positive rate. This N_COMI threshold is used to eliminate uninteresting rules mined from the actual transaction set.

N_IC and N_COMI are both necessary because they capture different properties of the rules. N_IC represents the specificity of terms in the rules while N_COMI captures the information shared by the antecedent and consequent. Our goal is to mine rules with highly informative terms where the rule mutual information is also high. The dual application of N_IC and N_COMI thresholds removes terms with little information and leads to the discovery of rules with high mutual information content.

3.4 Properties of Cross-ontology Mutual Information (N_COMI) and Integrated Rule Information Content ($IRIC$)

The N_IC of a concept t is 0 (lowest) when $p(t) = 1$ and is 1 (highest) when t occurs only once in the transaction dataset. The N_COMI of a rule is 0 when the concepts in the rule are statistically independent. The $IRIC$ of a rule is 0 (lowest) when the IC of both the concepts in the rule and the N_COMI of the rule are 0. The $IRIC$ is 1 (highest) when the IC of both the concepts in the rule and the N_COMI of the rule are 1. Tan et al. identify three key properties of a desirable metric ([16]): An interestingness metric, M is considered desirable if it satisfies the following three properties for a rule of the form $x \rightarrow y$.

1) M is 0 when x and y are statistically independent.
2) M monotonically increases with $p(xy)$ when $p(x)$ and $p(y)$ remain the same. $p(x)$, $p(y)$, and $p(xy)$ are the probabilities of observing x, y, or both in a transaction respectively.
3) M monotonically increases with $p(x)$ or $p(y)$ when the rest of the parameters remain the same.

These properties are meant to be applicable for metrics that quantify association rules and not individual terms. In our analysis, the metrics we use to quantify association rules are N_COMI and $IRIC$ while N_IC is used to quantify informativeness of terms. We list the behavior of N_COMI and $IRIC$ with respect to these properties in Table 1.

4. Main Results

We designed an experiment as a preliminary proof of concept to demonstrate the mining and ranking of cross-ontology relationships using the $IRIC$ metric. The data used for this experiment was gene expression data in post-natal mouse from the Gene Expression Database (GXD) [17] at the Mouse Genome Informatics (MGI). The transaction set built from this data contains 8,176 transactions and 123,069 GO terms and 124,920 anatomy terms. Each transaction contains a gene product name accompanied by one or more annotations to the anatomy and gene ontologies.

Cross-ontology rules were mined after generalization and the N_IC and N_COMI information theoretic metric thresholds were applied incrementally. The $IRIC$ metric, a combination of N_COMI and N_IC was used to rank the remaining mined rules after the thresholds were applied. In this experiment, we weighted GO and Anatomy concepts equally by setting α and β to 0.5.

4.1 Effect of N_IC and N_COMI thresholds

For this experiment, we chose to only include GO and Anatomy terms that were annotated to no more than 5% of the total genes in the dataset. This threshold was selected empirically. This translates into an IC of 4.32 (same for any dataset) and N_IC of 0.33 (specific to our dataset). The percentage of annotated genes in computing the N_IC threshold can be varied depending on the level of informativeness desired in the relationships. The greater the percentage of genes annotated to a term, the lower the N_IC of the term. A practical consideration in choosing this threshold is to explore the distribution of N_IC scores in the data and select a threshold that balances the number of terms for analysis and the information content of the terms. Different choices of percent genes annotated to a term and how this translates to IC (dataset independent) and N_IC (specific to our study) are shown in Table 2. The Monte Carlo method described in Section 3.3 was used to select a threshold for N_COMI. The selected N_COMI threshold was used to remove uninformative rules.

Table 3 provides a summary of the experimental results and shows the effect of N_IC and N_COMI in the mining process. We measure the effect of each of these components with respect to the number of rules mined, the average N_IC, and the average N_COMI.

When an N_IC threshold was applied alone, (Table 3, column 3), 91.16% of the mined rules are removed as uninteresting, the average N_IC increases by approximately 96% and the average N_COMI increases by approximately 55%.

When the N_COMI threshold is applied alone (without the N_IC threshold, Table 3 column 4) there is a much smaller (2%) reduction in the number of rules than seen with the N_IC cutoff. The average N_COMI of rules increases by 3.44%.

The last column in Table 3 demonstrates the synergistic effects of using both N_IC and N_COMI thresholds. Both the average N_IC and N_COMI scores are at the highest when both thresholds are applied together as compared to singular application of either one of the thresholds. These results demonstrate that the combined application of N_IC and N_COMI thresholds removes uninteresting rules effectively resulting in rules with high mutual information and containing informative terms.

4.2 Evaluation of the $IRIC$ metric

The $IRIC$ metric was evaluated by comparing it to a commonly used information theoretic measure, Information Gain [18]. The top 100 rules ranked by $IRIC$ were manually compared by a biologist to those ranked by Information Gain for evidence in published literature. The biologist attempted to validate each rule by conducting literature searches for evidence of a relationship between the antecedent and the consequent of the rule. If evidence of such a relationship was found in literature, the corresponding rule was categorized as "Validated" and the provenance information of the literature was recorded. If no evidence was found, the rule was marked as "Not Validated". Additionally, each rule was evaluated for its meaningfulness and categorized as either "Meaningful" or "Not meaningful". The meaningfulness of a rule indicates whether or not it makes sense for the items in the rule to be co-annotated [2]. The definitions of these categorizations are as per our previous work on cross-ontology association rule mining [2]. This evaluation was conducted based upon the biologist's personal, biological knowledge, and literature searches.

Of the top 100 $IRIC$ rules, literature-based evidence was found for 92 rules (S1 File). In contrast, evidence was found for only 78 of the top Information Gain rules. One $IRIC$ rule was categorized as "Not meaningful" while six Information Gain rules were categorized as "Not meaningful". These preliminary results show that $IRIC$'s top ranked rules have a high accuracy rate and that $IRIC$ outperforms Information Gain in the percentage of manually validated and meaningful rules. These evaluated rules and the results of manual validation are available publicly at (`https://github.com/prashanti/Supplementary_Files`).

5. Conclusions

The widespread use of ontologies to represent data and knowledge has led to the availability of vast amounts of ontology-annotated data. However, there is a dearth of efficient algorithms and ontology-aware metrics to mine multi-ontology data and rank the mined relationships. In this study, we developed information theoretic metrics to rank cross-ontology rules. We presented the use of our mining algorithm along with the metrics to mine relationships across the Gene Ontology and Anatomy Ontology. Our results demonstrate that our proposed metric, $IRIC$ is effective at ranking accurate relationships mined from annotation data.

5.1 Tables

Table 1: Properties satisfied by the information theoretic measures Cross-ontology Mutual Information (N_COMI) and Integrated Rule Information Content ($IRIC$).

Property	N_COMI	$IRIC$
1	Satisfies	Satisfies
2	Satisfies	Satisfies
3	Does not satisfy	Does not satisfy

Table 2: IC and N_IC thresholds corresponding to percentage of gene annotations to terms.

Threshold of % genes annotated to a term.	IC	N_IC (Number of genes = 8176)
25%	2.25	0.17
20%	2.32	0.18
15%	2.73	0.21
10%	3.32	0.26
5%	4.32	0.33
4%	4.64	0.36
3%	5.05	0.39
2%	5.64	0.43
1%	6.64	0.51

Table 3: Comparison of the number of rules mined, average N_IC, and average N_COMI when N_IC and N_COMI thresholds are applied individually and together.

	Before pruning	Only N_IC threshold applied	Only N_COMI threshold applied	Both N_IC and N_COMI thresholds applied
Number of rules mined	66437	5873	64908	4925
Average N_IC	0.28	0.55	0.33	0.55
Average N_COMI	0.058	0.13	0.06	0.148

References

[1] P. Manda, F. Mccarthy, and S. M. Bridges, "Interestingness measures and strategies for mining multi-ontology multi-level association rules from gene ontology annotations for the discovery of new go relationships," *J. of Biomedical Informatics*, vol. 46, no. 5, pp. 849–856, Oct. 2013. [Online]. Available: http://dx.doi.org/10.1016/j.jbi.2013.06.012

[2] P. Manda, S. Ozkan, H. Wang, F. McCarthy, and S. Bridges, "Cross-Ontology Multi-level Association Rule Mining in the Gene Ontology," *PLOS One*, 2012.

[3] G. Alterovitz, M. Xiang, M. Mohan, and M. Ramoni, "Go pad: the gene ontology partition database," *Nucleic Acids Research*, pp. 322–327, 2007.

[4] G. Alterovitz, M. Xiang, and M. Ramoni, "An information theoretic framework for ontology-based bioinformatics," in *Information Theory and Applications Workshop, 2007*, 29 2007-feb. 2 2007, pp. 16 –19.

[5] M. Ringwald, J. T. Eppig, D. A. Begley, J. P. Corradi, I. J. McCright, T. F. Hayamizu, D. P. Hill, J. A. Kadin, and J. E. Richardson, "The mouse gene expression database (gxd)," *Nucleic Acids Research*, vol. 29, no. 1, pp. 98–101, 2001. [Online]. Available: http://nar.oxfordjournals.org/content/29/1/98.abstract

[6] D. Faria, A. Schlicker, C. Pesquita, H. Bastos, A. E. Ferreira, M. Albrecht, and A. O. Falcão, "Mining go annotations for improving annotation consistency," *PloS one*, vol. 7, no. 7, p. e40519, 2012.

[7] P. Carmona-Saez, M. Chagoyen, A. Rodriguez, O. Trelles, J. M. Carazo, and A. Pascual-Montano, "Integrated analysis of gene expression by Association Rules Discovery," *BMC Bioinformatics*, vol. 7, p. 54, 2006.

[8] S. Myhre, H. Tveit, T. Mollestad, and A. Laegreid, "Additional gene ontology structure for improved biological reasoning," *Bioinformatics*, vol. 22, pp. 2020–2027, Aug 2006.

[9] O. Bodenreider, M. Aubry, and A. Burgun, "Non-lexical approaches to identifying associative relations in the gene ontology," *Pac Symp Biocomput*, pp. 91–102, 2005.

[10] F. Benites, S. Simon, and E. Sapozhnikova, "Mining rare associations between biological ontologies," *PloS one*, vol. 9, no. 1, p. e84475, 2014.

[11] I. N. Ferraz and A. C. B. Garcia, "Ontology in association rules," *SpringerPlus*, vol. 2, no. 1, p. 452, 2013.

[12] R. Paul, T. Groza, J. Hunter, and A. Zankl, "Semantic interestingness measures for discovering association rules in the skeletal dysplasia domain." *J. Biomedical Semantics*, vol. 5, p. 8, 2014.

[13] C. Borgelt and R. Kruse, "Induction of association rules: Apriori implementation," in *Proceedings of the 15th Conference on Computational Statistics*, 2002.

[14] C. E. Shannon, "A mathematical theory of communication," *The Bell system technical journal*, vol. 27, pp. 379–423, July 1948.

[15] Y. Ke, J. Cheng, and W. Ng, "An information-theoretic approach to quantitative association rule mining," *Knowl. Inf. Syst.*, vol. 16, no. 2, pp. 213–244, July 2008. [Online]. Available: http://dx.doi.org/10.1007/s10115-007-0104-4

[16] P.-N. Tan, V. Kumar, and J. Srivastava, "Selecting the right interestingness measure for association patterns," in *Proceedings of the eighth ACM SIGKDD international conference on Knowledge discovery and data mining*, ser. KDD '02. New York, NY, USA: ACM, 2002, pp. 32–41. [Online]. Available: http://doi.acm.org/10.1145/775047.775053

[17] J. H. Finger, C. M. Smith, T. F. Hayamizu, I. J. McCright, J. T. Eppig, J. A. Kadin, J. E. Richardson, and M. Ringwald, "The mouse Gene Expression Database (GXD): 2011 update," *Nucleic Acids Res.*, vol. 39, no. Database issue, pp. D835–841, Jan 2011.

[18] P. Lenca, P. Meyer, B. Vaillant, and S. Lallich, "On selecting interestingness measures for association rules: User oriented description and multiple criteria decision aid," *European Journal of Operational Research*, vol. 184, no. 2, pp. 610–626, 2008.

Inferring Gene Network From Gene Expression Data Using Dynamic Bayesian Network With Bayesian Optimization Algorithm and Different Scoring Metric Approaches

Muhammad Mahfuz Zainuddin[1], Mohd Saberi Mohamad[2], Lian En Chai[3], Zuraini Ali Shah[4], Weng Howe Chan[5], Safaai Deris[6], Hussah AlEisa[7], Saad Subair[8]

[1,2,3,4,5]Artificial Intelligence and Bioinformatics Research Group,Faculty of Computing, Universiti Teknologi Malaysia, 81310 Skudai, Johor, Malaysia

[6]Faculty of Creative Technology & Heritage, Universiti Malaysia Kelantan, Locked Bag 01, 16300 Bachok, Kota Bahru, Kelantan, Malaysia.

[7,8]Department of Computer Sciences, College of Computer and Information Sciences, Princess Nourah bint Abdulrahman University, Riyadh, KSA

Abstract- *A gene network can be referred as a model that uses directed graph which represents the regulation between genes. An inferring gene network can be defined as a process that identify the relationship or interaction between genes from the experiment data through the computational analysis. Our objectives in this research are to inferring gene network from gene expression data using Dynamic Bayesian Network with Bayesian optimization algorithm and different scoring metric approaches. In order to improve the gene network result by using Dynamic Bayesian Network, this research will implement Dynamic Bayesian Network with Bayesian Optimization algorithm to achieve it. To analyze the gene network, different scoring metric approaches that are BDe and MDL was used. This research uses microarray data that taken from Saccharomyces cerevisiae database. The result of this research is compared to a previous research to ensure the accuracy and validity of the results. When implementing the proposed algorithms and methods, this research is able to clarify the improvement of Dynamic Bayesian Network by using Bayesian Optimization algorithm and the effect of different scoring metric approaches to identify the biological relationship gene network within the gene expression datasets and then display the result in a suitable representation.*

Keywords: Gene network; Gene expressions; Bayesian network; Bayesian optimization algorithm; Scoring metric.

1 Introduction

Dynamic Bayesian Network (DBN) can be defined as a Bayesian Network that represents sequences of variables. Using time delay information, Dynamic Bayesian Network (DBN) can construct cyclic regulations. It uses the time series data to generate a relationship among random variables. DBN is an expansion version of Bayesian Network. It is a model that improves Bayesian network model. Inferring gene network can be translated as the process of identifying gene interaction from experiment data through computational analysis. Many researchers discussed this area of inferring gene network from gene expression data using dynamic Bayesian network and some other algorithms [8,9,10,11]. The data used in this purpose is gene expression data from microarray data. The aim is to infer gene network from gene expression data using Dynamic Bayesian Network with Bayesian Optimization and different scoring metric approaches. The scoring metric approaches that are used in this research are Bayesian Dirichlet comparability (BDe) and Minimum Description Length (MDL). To visualize the gene network, this research implemented Cytoscape software to conduct the research. Also, this research uses *Saccharomyces cerevisiae* gene expression data to evaluate the efficiency of the proposed method.

2 Material and Methods

In this research, we will describe the detail how DBN with BOA used for inferring GRNs from gene expression data. It consists three major step: missing value imputation, the construction of gene network, and evaluating the network structure using scoring metric approaches. The following sub-section will explain more about the three major steps.

2.1 Experimental Data and Missing Values Imputation

The data that being used here is *S. cerevisiae* cell cycle time-series gene expression data. This datasets contains missing values that must be processed. These missing values throw great impact on the modelling gene regulatory network result such as inaccurate predicted relationship among genes. One of the major steps in modelling of gene regulatory network is missing value imputation. The chosen imputation algorithm for this research is k-nearest neighbor (kNN) method. The k-nearest neighbor method is considered the simplest imputation algorithm. K-nearest neighbor imputation algorithm classifies the objects based on the closest training examples in the feature space. This algorithm is suited when there is few or no prior knowledge about the data distribution. According to [1], k-nearest neighbor algorithm provides more accurate and robust performance compare to several imputation algorithms. In addition, kNN-based imputation presents low level of deterioration in performance corresponding to the increasing percentage of missing values. Due to those reasons, k-nearest neighbor imputation algorithm is chosen for the estimation of missing values.

K-nearest neighbor imputation algorithm can be run in the MATLAB environment. The input dataset which have to impute missing values is imported to the MATLAB workspace at the first place. NaNs in the original dataset is meant for the missing values.

The algorithm obtains the value from the corresponding nearest-neighbor column which required for the imputation and replaces the obtained value to NaN. The nearest-neighbor column is the closest column with the target column which determined by using Euclidean distance function. The Euclidean distance between two points x = $(x_1,\ldots,x_n)$ and $y=(y_1,\ldots,y_n)$ in Euclidean n-space can defined as

$$d_E(x,y) = \sum_{i=1}^{n} \sqrt{x_i^2 - y_i^2} \qquad (1)$$

There is a condition that the corresponding value of the nearest-neighbor column also is NaN which meant for missing value. If this condition occurs, the value of next nearest column is used for imputation.

2.2 Learning Dynamic Bayesian Network using Bayesian Optimization Algorithm

After the imputation and discretization process, the sample is run through software called sBOA [2] in order to construct a Dynamic Bayesian Network and in order to learn Dynamic Bayesian Network that was constructed, we will use certain command in sBOA software where it can learn the network that constructed. According to [3], the Bayesian Optimization process involve four steps:

1. Establishing a fine solution group (using all kind of choice mechanisms to select solution group from the current population).
2. Find Dynamic Bayesian Network to matching solution group (using BD metric to find better network structure according to solution group).
3. Producing a new choice solution (using joint probability distributing of network diagram to produce new individual groups).
4. Creating next population (using new individual group to replace elderly group and update population as a newer population).
5. Repeat until meet termination rule (good solution has been found).

Figure1 shows a diagram that illustrates the learning Dynamic Bayesian Network using Bayerian Optimization Algorithm process and method.

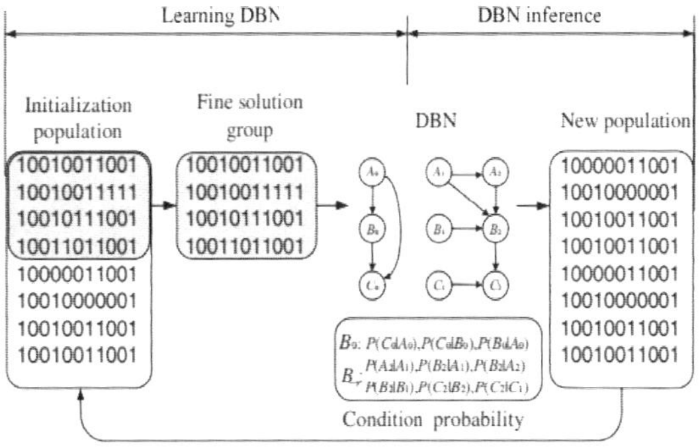

Figure 1: Diagram on learning Dynamic Bayesian Network using Bayerian Optimization Algorithm.

2.3 Constructing Gene Network

To construct gene network, this research use Cytoscape software. This software is open source bioinformatics software that can visualize molecular interaction network and biological pathways and integrating the network with other state data. This software was used to generate a graph that represents the DBN-BOA for *Saccharomyces cerevisiae* datasets. After gene networks are constructed using dynamic Bayesian Network with BOA, it is evaluated in term of performance. The network constructed are compared with that used in [4] which is illustrated in Appendix A at the end of this paper.

The comparison of the performance is conducted by calculating the True Positive(TP), False Positive (FP), True Negative (TN), and False Negative (FN) edges. True positive (TP) is the number of edges that exist in the network constructed by [4] and in the network formed in this research. The False Positive(FP) is the number of the edges that do not exist in the network done by [4] but exist in this research. True Negative (TN) is the number of edges that do neither exist in this network nor in [4]. False Negative (FN) is the number of edges that exist in the network of [4] but not exist in the network of this research.

2.4 Evaluating Network Structures

After gene system have been developed and diagram been produced, then we evaluated the network structure by scoring metric. In this research, two distinctive scoring metric were applied in order to come with good network structure; namely Bayesian Dirichlet comparability (BDe) and Minimum Description Length (MDL). All of these scores attempt to balance the fitness of the proposed model to the data with the model's complexity. The BDe scores decompose into aggregation of scores of each of the BN's families. BDe are also known for their better solution quality whereas MDL are good in speed.

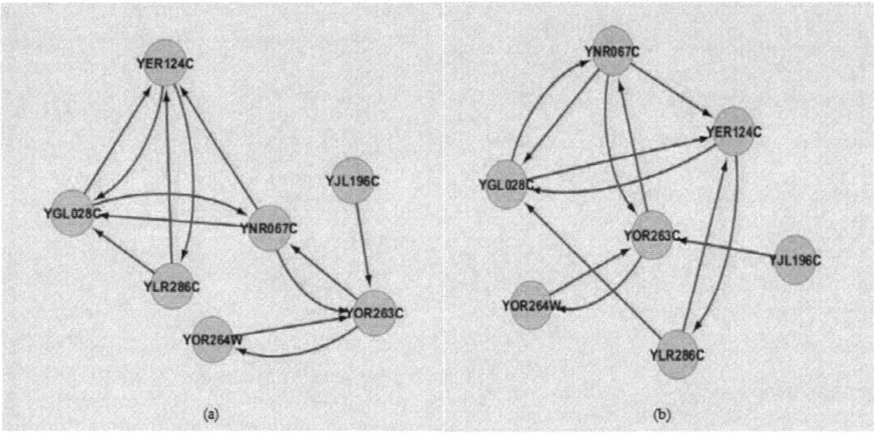

Figure 2: YOR263C sub-network constructed by DBN-BOA with (a) BDe and (b) MDL.

3 Results and Discussion

In this work, we will compare our result with [4] result. The sub-network that chosen from [4] are YOR263C sub-network and YPL256C sub-network.

3.1 YOR263C Sub-Network

Figure 2(a) and Figure 2(b) show the YOR263C sub-network that is constructed in this research using BDe and MDL scoring metric approaches. It is very clear that both networks consist of 7 nodes and 13 edges. It shows a different number of edges from that obtained in [4]. Through this research, we can see that several edges in the network are form cyclic regulation and have at least one directed edge with other nodes, while the network constructed by [4] does not show any cyclic regulation. The main difference between this research and [4] is that the interactions between genes are clearer. As we can see in the sub-network from [4], the edge formed between YOR263C and YOR264W cannot show which gene is regulating which. However, this research shows that YOR 263C is regulating YOR264W and it is a cyclic regulation. This means that the expression level of YOR264W depends on both YOR263C and YNR067C. Here, three new edges were identified. However, it is shown that Dynamic Bayesian networks with Bayesian Optimization algorithm implemented in this research are able to construct cyclic regulation and form more potential edges between genes in a sub-network.

Table 1 shows the comparison of edges in YOR263C sub-network among network constructed by [4] and this research. True Positive (TP) is the number of edges that exist in both network constructed by [4] and this research. False Negative (FN) is the number of edges that exist in the sub-network of [4], but does not exist in network of this research. False Positive (FP) is the number of edges that exist in network of this research, but does not exist in [4]. True Negative (TN) is the number of edges that does not exist in both network constructed by [4] and in this research.

The sensitivity (true positive rate) for YOR263C sub-network is approximately 83.33% by five edges that exist in [4] and also exist in the network of this research. There are about five cyclic edges were formed in this subnetwork. This shows that dynamic Bayesian networks implemented in this research has successfully predicted and formed all the possible existing edges or interactions between genes in the network almost like [4] even they have cyclic regulation in the sub-network.

Table 1: Comparison of YOR263C Sub-Networks

Condition	Number of Edges	Statistical Performance
True Positive (TP)	5	Sensitivity TP/(TP+FN) = 5/(5+1) = 83.33%
False Negative (FN)	1	
False Positive (FP)	3	Specificity TN/(FP+TN) = 12/(3+12) = 80.00%
True Negative (TN)	12	

Table 2: Comparison Between Basic DBN and DBN-BOA Based On YOR263C Sub-Network

Condition	DBN	DBN-BOA
True Positive (TP)	4	5
False Negative (FN)	2	1
False Positive (FP)	2	3
True Negative (TN)	12	12
Sensitivity	66.67%	83.33%
Specifity	85.71%	80.00%
Accuracy	80.00%	80.95%

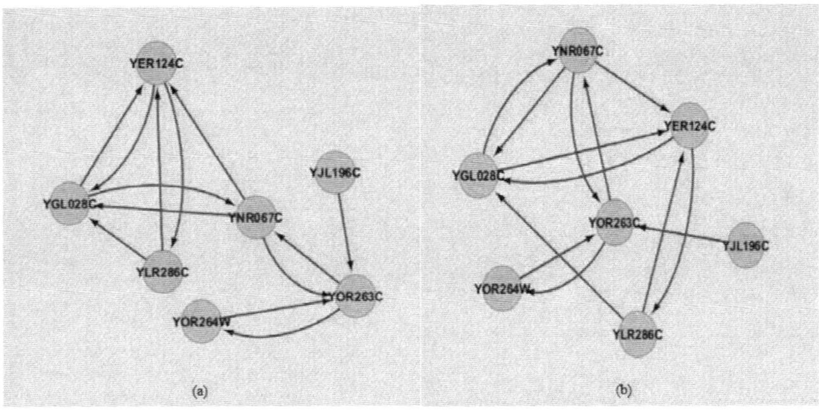

Figure 3: YPL256C sub-network constructed by DBN-BOA with (a) BDe and (b) MDL.

The specificity (true negative rate) for this sub-network is approximately 80%, whereby there are three edges formed in this network but does not exist in the sub-network in [4]. The three edges are YNR067C to YER124C andYOR263C, and YJL196C with YOR263C. This shows that the Dynamic Bayesian networks with Bayesian Optimization algorithm implemented in this research is capable of uncovering more potential edges, interactions and cyclic regulation between genes compared with [4].

Table 2 shows the comparison between DBN-BOA with basic DBN. It shows that in term of sensitivity, basic DBN get 66.67% and the proposed method (DBN-BOA) show 83.33%. For the Specificity, basic DBN shows 85.71 and DBN-BOA show 80 %. In term of accuracy, the proposed method showed slightly higher accuracy than the basic DBN which is 80 .95%. Thus, it can be concluded that DBN-BOA produce more accurate result.

3.2 YPL256C sub-Network

Figure3(a) and Figure3(b) shows the YPL256C sub-network that is constructed in this research using BDe and MDL scoring metric approaches. It is shown clear that both networks consist of 12 nodes and 24 edges. It shows a different number of edges obtain in this research as compared to [4]. Through this research, we can see that several edges in the network are form cyclic regulation and have at least one directed edge with other nodes. While the network done by [4] does not show cyclic regulation and the gene YGR108W failed to construct with any edge. About 20 new edges were identified in this research which is twice the edges compared to [4]. Hence, it proven that dynamic Bayesian networks implemented in this research are able to construct cyclic regulation and form more potential edges between genes in a sub-network.

Table 3 shows the comparison of edges formed in YPL256C sub-network of [4] and this research. True Positive (TP) is the number of edges that exist in both networks constructed by [4] and this research. False Negative (FN) is the number of edges that exist in [4] sub-network, but does not exist in network of this research. False Positive (FP) is the number of edges that exist in network of this research, but does not exist in [4]. True Negative (TN) is the number of edges that does not exist in both networks that are constructed by [4] and in this research.

Table 3: Comparison Of YPL256C Sub-Networks

Condition	Number of Edges	Statistical Performance
True Positive (TP)	4	Sensitivity TP/(TP+FN) = 4/(4+5) = 44.44%
False Negative (FN)	5	
False Positive (FP)	20	Specificity TN/(FP+TN) = 113/(20+113) = 84.96%
True Negative (TN)	113	

The sensitivity (true positive rate) for this sub-network is 44% formed by four directed edges that exist in the network of [4] which are also been captured in this research as well. However, there are about five directed edges exists in [4] but it does not exist in network of this research. The missing edge is between gene YPL256C to YIL140W and YIL066C, gene YER001W to YPL256C, gene YMR001C to YLR131C and YDR146C respectively.

The specificity (true negative rate) for this sub-network is approximately 84.96%. There are about 20 new edges between nodes that were unable to be captured by [4]. The new edges are YPL163C to YIL140W and YMR199W, YIL140W to YER001W and YMR199W, YER001W to YIL140W and YMR199W, YMR199W to YPL256C, YPL256C to YGL021, YIL066C to YMR199W and YGL021W, YHR023W to YDR146C and YGL021W, YDR146C to YGL021, YLR131C to YMR001C and YGL021W, YMR001C to YGL021W, YGR108W to YMR001C and YGL021W (cyclic respectively). There are about seven cyclic edges were formed in this sub-network. Gene YLR131C regulates both gene YMR001C and YGL021W. Gene YLR131C encodes for transcription factor that activates transcription of genes expressed in the G1 phase of the cell cycle. On the other hand, gene YMR001C involved in regulation of DNA replication which encodes protein. Furthermore, gene YIL066C is expressed only after DNA damage occurred in order to match with the function of YMR199W. Gene YMR199W encodes for G1-cyclin which involved in regulation of the cell cycle. Therefore, it is biologically logical for YIL066C to regulate the expression of YMR199W.

Table 4: Comparison Between Basic DBN And DBN-BOA Based On YPL256C Sub-Network

Condition	DBN	DBN-BOA
True Positive (TP)	3	4
False Negative (FN)	6	5
False Positive (FP)	22	20
True Negative (TN)	118	113
Sensitivity	33.33%	44.44%
Specifity	84.28%	84.96%
Accuracy	81.21%	82.39%

Table 4 shows the comparison between DBN-BOA with basic DBN for YPL256C sub-network. In term of sensitivity, basic DBN get 33.33% and the proposed method (DBN-BOA) show 44.44%. For the Specificity, basic DBN shows 84.28% and DBN-BOA show 84.96%. In terms of accuracy, our proposed method shows slightly higher result (82.39%) than the basic DBN (81.21%). It can be concluded that DBN-BOA produce more accurate result.

3.3 Performance Scoring Metric

Table 5 shows a comparison of YOR263C sub-networks based on scoring metric approaches. BDe are known for the better solution quality whereas MDL are good in speed. According to [5], BDe scoring metric is very times demanding, a single number on a dataset of 20 genes and 300 observations can take up to a day, while MDL scoring metric is very fast. As shown in the table above, both computation times taken by BDe and MDL are about one second only and the results obtain from this network are the same. Both networks with BDe and MDL scoring metric give 13 numbers of edges and 7 numbers of nodes. The numbers (TP, FN, FP, TN) for both network are also the same as shown in the Table 5. Further, [6] also showed that BDe seems to learn more accurate networks than MDL (which is also equivalent to the BIC). It also shows that it has the same computational time with the basic Dynamic Bayesian network.

Table 6 shows a comparison of YPL256C sub-networks with scoring metric approaches. BDe are known for the better solution quality whereas MDL are good in speed. According to [5], BDe is very times demanding, a single number on a dataset of 20 genes and 300 observations can take up to a day, while MDL is very fast. As shown in the table above, computation times taken by BDe are about two minutes compared to MDL that took one minute only. But the result obtain from this network are same. Both networks with BDe and MDL scoring metric give 24 numbers of edges and 12 numbers of nodes. The condition (TP, FN, FP, TN) for both network are also same as shown in Table 6. The BDe scoring metric is still recommended over MDL scoring metric, that is due to its easiness in the statistical interpretation [7]. From this research, it shows that the computational time for Dynamic Bayesian Network with Bayesian Optimization algorithm is faster than basic Bayesian Network.

Table 5: Comparison Of Yor263c Sub-Networks Based On Computational Time

Scoring Metric	DBN (HH:MM:SS)	DBN-BOA (HH:MM:SS)
BDe	00:00:01	00:00:01
MDL	00:00:01	00:00:01

Table 6: Comparison Of Ypl256c Sub-Networks Based On Computational Time

Scoring Metric	DBN (HH:MM:SS)	DBN-BOA (HH:MM:SS)
BDe	00:02:01	00:01:58
MDL	00:01:10	00:01:02

4 Conclusion

In this research, our aim is to improve the accuracy of basic DBN gene network result using enhanced Dynamic Bayesian Network with Bayesian Optimization Algorithm. From the results, it is shown that it achieved better performance than the Bayesian Network conducted by [4] and basic Dynamic Bayesian Network. It was concluded that DBN-BOA is able to improve the accuracy for prediction and reveal more of the novel potential interactions between genes compared to [4] and basic Dynamic Bayesian Network. Also, in this research, BDe and MDL scoring metric was applied to DBN-BOA model to get the optimal network structure. From the result, it is shown that the MDL scoring metric has faster computation speed for large networks but in term of accuracy, BDe has exactness in the statistical interpretation of the network. Thus, in order to find an accurate network, we suggest using BDe scoring metric and if required less computational time, use MDL scoring metric. For the future work, we plan to apply other optimization algorithm and different scoring metric to improve network accuracy.

5 Acknowledgments

We would like to thank Universiti Teknologi Malaysia for funding this research through the Research University Grant (Grant number: Q.J130000.2528.12H12). This research also supported by the Malaysian Ministry of Education through the Fundamental Research Grant Scheme (Grant number: R.J130000.7828.4F720). Acknowledgements are also extended to the other contributing persons.

6 References

[1] O. Troyanskaya, M. Cantor, G. Sherlock, P. Brown, T. Hastie, R. Tibshirani, D. Botstein, and R.B. Altman, "Missing value estimation methods for DNA microarrays," Bioinformatics, vol. 17, no. 6, pp. 520-525, June 2001.

[2] M. Pelikan, "Bayesian optimization algorithm," in Hierarchical Bayesian Optimization Algorithm, Springer Berlin Heidelberg, 2005, pp. 31-48.

[3] S. Gao, Q. Xiao, Q. Pan, and Q. Li, "Learning dynamic Bayesian networks structure based on Bayesian optimization algorithm," in Advances in Neural Networks

– ISNN 2007, D. Liu, S. Fei, Z. Hou, H. Zhang, and C. Sun, Eds. Springer Berlin Heidelberg, 2007, pp. 424-431.

[4] D. Mathaus, "Analyzing gene expression data with bayesian networks," M.S. thesis, Institute of Biomedical Engineering, Technische Universitat Graz, Graz, Austria, 2002.

[5] N.X. Vinh, M. Chetty, R. Coppel, and P.P. Wangikar, "GlobalMIT: Learning Globally Optimal Dynamic Bayesian Network with the Mutual Information Test (MIT) Criterion," Bioinformatics, vol. 27, no. 19, pp. 2765-2766, October 2011.

[6] L.M. de Campos, "A Scoring Function for Learning Bayesian Networks based on Mutual Information and Conditional Independence Tests," Journal of Machine Learning Research, vol. 7, pp. 2149-2187, October 2006.

[7] B. Wilczyński, and N. Dojer, "BNFinder: exact and efficient method for learning Bayesian networks," Bioinformatics, vol. 25, no. 2, pp. 286-287, Jauary 2009.

[8] Ram R, Chetty M. "A Markov-Blanket-Based model for gene regulatory network inference". IEEE/ACM Trans Comput Biol Bioinf. 8: 353-367, 2011.

[9] Greenfield A., et al. "Robust data-driven incorporation of prior knowledge into the inference of dynamic regulatory networks". Bioinformatics, 29, 1060–1067, 2013.

[10] Li Y, Jackson SA. "Gene Network Reconstruction by Integration of Prior Biological Knowledge". G3 (Bethesda) 5(6):1075-9. doi: 10.1534/g3.115.018127, 2015

[11] Oh JH, Deasy JO. "Inference of radio-responsive gene regulatory networks using the graphical lasso algorithm". BMC Bioinformatics., 15 Suppl 7:S5. doi: 10.1186/1471-2105-15-S7-S5, 2014

Appendix A

Figure shows YPL256C sub-network constructed by DBN-BOA as reported by [4]

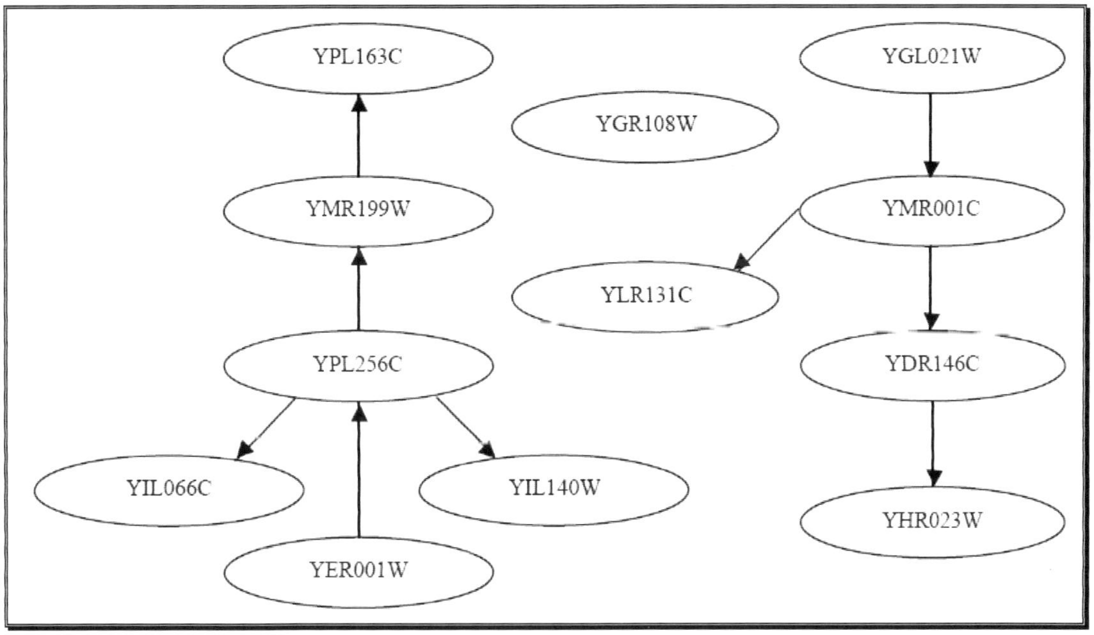

Increasing Efficiency of Microarray Analysis by PCA and Machine Learning Methods

Jing Sun[1], **Kalpdrum Passi**[1], **Chakresh Jain**[2]
[1]Department of Mathematics and Computer Science, Laurentian University, Sudbury, Ontario, Canada
[2]Department of Biotechnology, Jaypee Institute of Information Technology, Noida, India

Abstract - *Principal Component Analysis (PCA) is widely used method for dimensionality reduction. However, it has not been studied much as a feature selection method to increase the efficiency of the classifiers on microarray data analysis. In this study, we assessed the performance of four classifiers on the microarray datasets of colon and leukemia cancer before and after applying PCA as a feature selection method. Different thresholds were used with 10-fold cross validation. Significant improvement was observed in the performance of the well-known machine learning classifiers on microarray datasets of colon and leukemia after applying PCA.*

Keywords - Principal component analysis; Support Vector Machine; Random Forest; Neural Network; K-Nearest-Neighbor; Feature selection.

1 Introduction

The gene expression profiling techniques by DNA microarrays provide the analysis of large amount of genes [1]. The amount of gene expression data of microarray has grown exponentially. It is of great importance to find the key gene expression which can best describe the phenotypic trait [2]. The microarray dataset usually has a large number of genes in small number of experiments which collectively raise the issue of "curse of dimensionality" [17]. To find the key gene expression, one way is to use feature selection methods. In this paper we use Principal Component Analysis (PCA) for feature selection and apply four well-known machine learning methods, Support Vector Machine (SVM), Neural Network (NN), K-Nearest-Neighbor (KNN) and Random Forest algorithms to validate and compare the performance of Principal Component Analysis. In the first set of experiments presented in this paper, the performance of the four machine learning techniques (SVM, NN, KNN, Random Forest) is compared on the colon and leukemia microarray datasets. The second set of experiments compares the performance of these machine learning algorithms by applying PCA method on the same datasets.

The main contribution of this research is to show that PCA used as a feature selection method can improve the performance of the four well known machine learning algorithms on microarray datasets.

The paper is organized as follows. We first state our experiment environment in Section II. Secondly, literature review is given in Section III. Methodology is explained in Section IV; results and discussion are given in Section V. Section VI presents the conclusions of this study and states future prospects and limitation of this work.

2 Dataset and Tools

2.1 Dataset

The datasets used in this comparison study are the "Colon cancer dataset" and the "Leukemia cancer dataset". They can be accessed from various sources. The colon dataset describes a colon cancer study [3] in which gene expression levels were measured on 40 normal tissues and 22 tumor tissues. The leukemia dataset consists of 38 bone marrow samples obtained from acute leukemia patients and 34 normal samples [4]. There are 2000 attributes corresponding to 2000 different genes for each tissue. The leukemia dataset has 7129 attributes corresponding to 7129 different genes for each tissue. The datasets were already normalized to mean zero and variance at the time of downloading.

2.2 Tools

In our experiment, we use Weka as the main testing tools which is a collection of machine learning algorithms for data mining tasks. All the results in this paper have been obtained from Weka 3-7-13-oracle-jvm on Mac OS 10.11.1. In Weka and R, we can access implementation of PCA from its library. Some of the plots and tables have been obtained from R. The version of R used in the experiments is 3.2.2.

3 Literature Review

Disease classification is the primary issue of microarray research. We can see from [5] that most of the previous analysis and reporting focused on outcome-related gene finding, class discovery and supervised prediction. Most of the studies focus on Hematologic malignancies, Lung and pleura [6][7], Breast [8][9], Hepato-digestive system, etc. As far as the authors are aware, few studies have been carried out that investigate the effect of using PCA with a number of machine learning algorithms.

Tom Howley [10] states the usefulness of PCA for reducing dimensionality and improving the performance of a variety of machine learning methods. Previous work mostly focuses on some specific machine learning algorithms [11][12].

PCA is a classical statistical method for transforming attributes of a dataset into a new dataset of uncorrelated attributes called principal components. PCA can be used as a dimensionality reduction method. The goal of this research is to determine if PCA can be used to improve the performance of machine learning algorithms in the classification of colon and leukemia datasets.

PCA is an exploratory multivariate statistical technique for simplifying complex data sets [13]. Given m observations on n variables, the goal of PCA is to reduce the dimensionality of the data matrix by finding r new variables, where r < n. Termed principal components, these r new variables together account for as much of the variance in the original n variables as possible while remaining mutually uncorrelated and orthogonal. Each principal component is a linear combination of the original variables, and so it is often possible to ascribe meaning to what the components represent. Principal Components Analysis has been used in a wide range of biomedical problems, including the analysis of microarray data in search of outlier genes [14] as well as the analysis of other types of expression data [15].

4 Methodology

There are three parts in our experiments: feature selection by PCA, cross-validation comparison and ratio comparison. We apply all the three parts on the colon and leukemia datasets.

4.1 Feature selection

Principal Component Analysis (PCA) is a multivariate technique that analyzes a data table in which observations are described by several inter-correlated quantitative dependent variables. Its goal is to extract the important information from the table, to represent it as a set of new orthogonal variables called principal components, and to display the pattern of similarity of the observations and of the variables as points in maps. The quality of the PCA model can be evaluated using cross-validation techniques. Mathematically, PCA depends upon the eigen-decomposition of positive semi-definite matrices and upon the singular value decomposition (SVD) of rectangular matrices [16]. The PCA viewpoint requires that one compute the eigenvalues and eigenvectors of the covariance matrix, which is the product XX^T, where X is the data matrix. Since the covariance matrix is symmetric, the matrix is diagonalizable, and the eigenvectors can be normalized such that they are orthonormal:

$$XX^T = WDW^T \qquad (1)$$

On the other hand, applying SVD to the data matrix X as follows:

$$X = U\Sigma V^T \qquad (2)$$

and attempting to construct the covariance matrix from this decomposition gives

$$XX^T = (U\Sigma V^T)(U\Sigma V^T)^T \qquad (3)$$

$$XX^T = (U\Sigma V^T)(V\Sigma U^T) \qquad (4)$$

and since V is an orthogonal matrix($V^TV = I$),

$$XX^T = U\Sigma^2 U^T \qquad (5)$$

and the correspondence is easily seen.

For each experiment, we need the original dataset and the new dataset obtained by applying the PCA. Proportion of variance is an important value in PCA which gives the main idea of how much variance this new attribute covered. Our selection uses this value to be the threshold and we choose different thresholds for selecting new subsets of data from the original one. We then obtain different datasets with threshold values of 95%, 90%, ..., 50%.

4.2 Cross-Validation in Principle Component Analysis

10-fold cross validation was applied for each classifier on all the datasets. The performance was compared for correctly classified instances and area under the ROC (Receiver Operating Characteristic) curve (AUC).

4.3 Ratio Comparison

The process of Ratio Comparison in PCA is that we split the dataset in different ratios of training set and test set. The four classifiers are applied on the datasets and use the test set to validate the model. The results are compared by correctly classified instances and AUC.

5 Results and Discussion

5.1 Principal Component Analysis Dataset List

We applied PCA on the colon and leukemia datasets. The variance table returned by PCA is listed in Table 1 and Table 2.

Table 1. Colon Dataset Thresholds and Attribute Selection

Colon Dataset Thresholds	Cumulative Proportion	Attributes Selected
100%(Raw)	100%	2001
95%	95.013%	45
90%	90.520%	35
85%	85.677%	27
80%	80.006%	20
75%	75.545%	16
70%	71.429%	13
65%	66.004%	10
60%	61.154%	8
55%	57.701%	7
50%	53.180%	6

The experiment is based on the 11 datasets shown in Table 1 and Table 2. The 100% dataset threshold means we use the raw data as input for the experiments. The 95% to 50% datasets are chosen by PCA method.

Table 2. Leukemia Dataset Thresholds and Attributes Selection

Dataset Thresholds	Cumulative Proportion	Attributes Selected
100%(Raw)	100%	7130
95%	95.192%	59
90%	90.244%	49
85%	85.560%	41
80%	80.232%	33
75%	75.638%	27
70%	70.261%	21
65%	65.997%	17
60%	60.570%	13
55%	55.557%	10
50%	51.440%	8

5.2 10-folds Cross Validation Results

The results are listed in Table 3. The accuracy (correctly classified instances) is given by:

$$accuracy = \frac{TP+TN}{N} \quad (6)$$

where TP indicates the True Positive instances, TN indicates the True Negative instances and N is the total number of instances in the test set.

Table 3 shows the accuracy and Area Under ROC curve (AUC) for the four classifiers on the colon dataset. K-nearest-neighbor and Random Forest algorithms shows the highest accuracy for a threshold of 60% by PCA. SVM algorithm shows the highest accuracy for the raw data and for 90% threshold by PCA. Neural Network algorithm shows the highest accuracy for a threshold of 90% by PCA. All the four classifiers show improvement in accuracy for some threshold value of PCA as compared to raw data except for SVM. Figure 1 shows the results of the 10-folds cross validation for colon dataset.

Table 3. 10-Folds Cross Validation For Colon Dataset

Data Mining Methods	Dataset	Accuracy	ROC Area(auc)
KNN	*Raw*	72.5806%	0.699
	95%	70.9677%	0.680
	90%	66.129%	0.648
	85%	70.9677%	0.728
	80%	64.5161%	0.619
	75%	62.9032%	0.581
	70%	62.9032%	0.592
	65%	70.9677%	0.706
	60%	82.2581%	0.815
	55%	75.8065%	0.752
	50%	72.5806%	0.744
Random Forest	*Raw*	82.2581%	0.885
	95%	69.3548%	0.879
	90%	74.1935%	0.877
	85%	77.4194%	0.840
	80%	74.1935%	0.812
	75%	77.4194%	0.845
	70%	79.0323%	0.855
	65%	82.2581%	0.873
	60%	85.4839%	0.892
	55%	82.2581%	0.881
	50%	82.2581%	0.872
SVM	*Raw*	87.0968%	0.886
	95%	79.0323%	0.868
	90%	82.2581%	0.893
	85%	74.1935%	0.805
	80%	70.9677%	0.797
	75%	66.129%	0.759
	70%	69.3548%	0.723
	65%	70.9677%	0.830
	60%	74.1935%	0.903
	55%	79.0323%	0.869
	50%	80.6452%	0.881
Neural Network	*Raw*	77.8%	0.857
	95%	79.0323%	0.851
	90%	83.871%	0.895
	85%	72.5806%	0.819
	80%	70.9677%	0.777
	75%	77.4194%	0.845
	70%	75.8065%	0.786
	65%	77.4194%	0.805
	60%	80.6452%	0.843
	55%	82.2581%	0.834
	50%	79.0323%	0.826

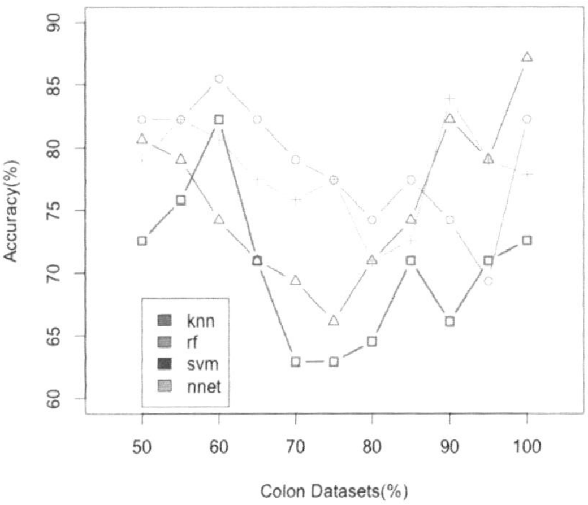

Figure 1. 10-fold cross validation results for colon dataset

Table 4 shows the accuracy and Area Under ROC curve (AUC) for the four classifiers on the leukemia dataset. K-nearest-neighbor shows the highest accuracy for a threshold of 70% by PCA. Random Forest shows the highest accuracy for a threshold of 65% and 70% by PCA. SVM shows the highest accuracy for the raw data and next highest accuracy for a threshold of 95% by PCA. Neural Network shows the highest accuracy for a threshold of 60% and 50% by PCA. Figure 2 shows the results of the 10-folds cross validation for leukemia dataset.

5.3 Ratio Validation Results

The second method we use for this experiment is that we split the dataset to a training set and test set by different ratio in 90%:10%,80%:20%,70%:30% and 60%:40%. All the result applied to the data which preprocessed by PCA. We show the results in Table 5 and Table 6.

Figures 3 to 6 show the accuracy for the four algorithms for different ratios of training and test datasets. Further discussion is given below.

Table 4. 10-Folds Cross Validation For Leukemia Dataset

Data Mining Methods	Dataset	accuracy	ROC Area (auc)
KNN	*Raw*	65.2778%	0.505
	95%	72.2222%	0.656
	90%	75%	0.667
	85%	77.7778%	0.719
	80%	81.9444%	0.811
	75%	83.3333%	0.829
	70%	93.0556%	0.911
	65%	91.6667%	0.887
	60%	88.8889%	0.861
	55%	90.2778%	0.874
	50%	91.6667%	0.874
Random Forest	*Raw*	76.3889%	0.889
	95%	79.1667%	0.918
	90%	84.7222%	0.945
	85%	84.7222%	0.952
	80%	93.0556%	0.963
	75%	93.0556%	0.958
	70%	94.4444%	0.978
	65%	94.4444%	0.974
	60%	91.6667%	0.963
	55%	90.2778%	0.968
	50%	91.6667%	0.969
SVM	*Raw*	98.6111%	0.998
	95%	97.2222%	0.995
	90%	86.1111%	0.969
	85%	87.5%	0.959
	80%	93.0556%	0.968
	75%	90.2778%	0.977
	70%	90.2778%	0.974
	65%	88.8889%	0.963
	60%	93.0556%	0.969
	55%	90.2778%	0.933
	50%	86.1111%	0.962
Neural Network	*Raw*	81.9444%	0.865
	95%	83.3333%	0.877
	90%	87.5%	0.917
	85%	90.2778%	0.934
	80%	90.2778%	0.970
	75%	90.2778%	0.977
	70%	88.8889%	0.980
	65%	88.8889%	0.971
	60%	93.0556%	0.971
	55%	91.6667%	0.951
	50%	93.0556%	0.974

5.4 Discussion

In Table 3 for the colon dataset, we observe that K-nearest-neighbor algorithm gives the highest accuracy of 82.3% for a threshold of 60% by PCA as compared to 72.6% for the raw data, an increase of 9.7% in accuracy by applying PCA. Random Forest algorithm gives the highest accuracy of 85.5% for a threshold of 60% by PCA as compared to 82.3% for the raw data, an increase of 3.3% in accuracy. Neural Network algorithm gives the highest accuracy of 83.9% for a threshold of 90% by PCA as compared to 77.8% for the raw data, an increase of 6.1% in accuracy. However, in SVM, the highest accuracy was observed as 87% for the raw data and 82% accuracy for a threshold of 90% by PCA, a decrease of 5% in accuracy.

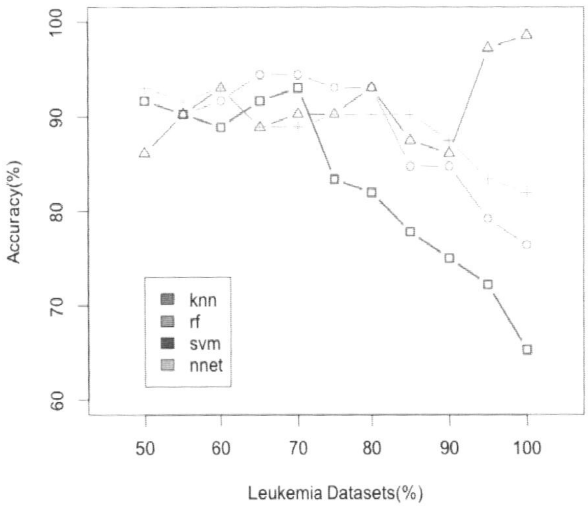

Figure 2. 10-fold cross validation results for leukemia dataset

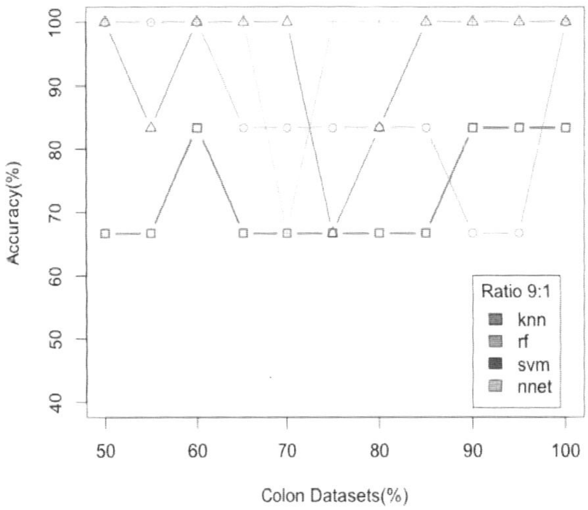

Figure 3. Accuracy comparison in Ratio 9:1

In Table 4 for the leukemia dataset, we observe that K-nearest-neighbor algorithm gives the highest accuracy of 93% for a threshold of 70% by PCA as compared to 65% for the raw data, an increase of 28% in accuracy by applying PCA. Random

Forest algorithm gives the highest accuracy of 94.4% for a threshold of 65% and 70% by PCA as compared to 76.4% for the raw data, an increase of 18% in accuracy. Neural Network algorithm gives the highest accuracy of 93% for a threshold of 50% and 60% by PCA as compared to 81.9% for the raw data, an increase of 11% in accuracy. However, in SVM, the highest accuracy was observed as 98.6% for the raw data as compared to 97.2% for a threshold of 95% by PCA, a decrease of 3.6% in accuracy.

SVM was tested for four different kernels – linear, polynomial, radial basis function and sigmoid function. The linear kernel gave the best results. For the exception of SVM, all other algorithms increased the accuracy of classification by applying PCA. Greater increase in accuracy was observed in leukemia dataset than the colon dataset. Figures 1 and 2 show the results for 10-fold cross validation for the colon and leukemia datasets, respectively.

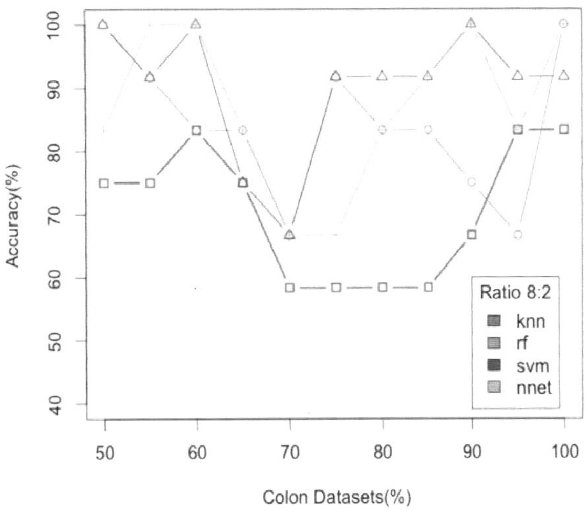

Figure 4. Accuracy comparison in ratio 8:2

Tables 5 and 6 show the accuracy and AUC of the colon and leukemia datasets respectively for raw data and different thresholds of PCA and by taking different ratios of training and test data.

In the colon dataset, we observe that highest accuracy is achieved for the training to test ratio of 9:1 and second highest accuracy for the ratio 8:2 for all the algorithms with and without using PCA.

In training to test ratio 9:1, Random Forest, SVM and Neural Network algorithms give an accuracy of 100% whereas k-nearest-neighbor gives an accuracy of 83.3% for the raw data. PCA maintains the accuracy of 100% at the threshold of 50% and 60% for Random Forest, SVM and Neural Network and maintains the accuracy of 83.3% at the threshold of 60%, 90% and 95% for the k-nearest-neighbor algorithm. Figure 3 shows the comparison of accuracy for the four algorithms for the ratio 9:1.

In training to test ratio 8:2, Random Forest and Neural Network algorithms give an accuracy of 100%, SVM has an accuracy of 91.6% and KNN has an accuracy of 83.3% for the raw data. PCA maintains the accuracy of 100% at thresholds of 55% and 60% for Neural Network and at the threshold of 50% for Random Forest. PCA increased the accuracy of SVM from 91.6% to 100% at the thresholds of 50%, 60% and 90%. PCA maintains the accuracy of KNN at 83.3% at a threshold of 60%. Figure 4 shows the comparison of accuracy for the four algorithms for the ratio 8:2.

In training to test ratio 7:3, PCA increases the accuracy of KNN from 78.95% to 84.21% at a threshold of 60% and increases the accuracy of Neural Network from 89.5% to 94.7% at a threshold of 90%. PCA maintains the accuracy of SVM at 89.5% at the thresholds of 50%, 90% and 95%. However, the accuracy of Random Forest is decreased from 89.5% to 84.2% at thresholds of 55% and 80%. Figure 5 shows the comparison of accuracy for the four algorithms for the ratio 7:3.

In training to test ratio of 6:4, PCA increases the accuracy of KNN from 76% to 84% at a threshold of 60%. PCA maintains the accuracy of SVM and Neural Network at 88% at a threshold of 55% for SVM and at 85% and 90% for Neural Network. However, the accuracy of Random Forest decreased from 88% to 76% at a threshold of 95%. Figure 6 shows the comparison of accuracy for the four algorithms for the ratio 6:4.

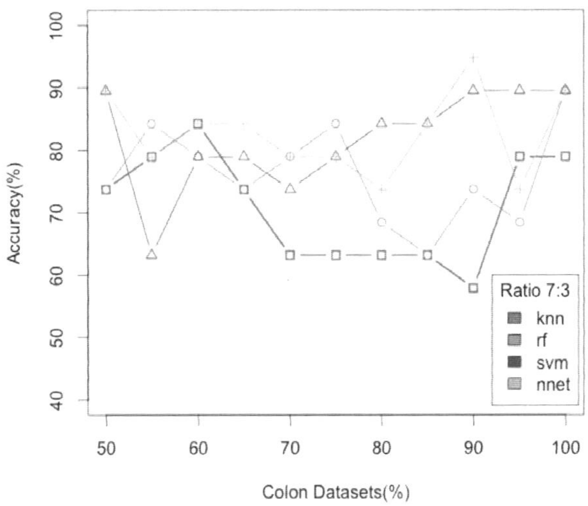

Figure 5. Accuracy comparison in Ratio 7:3

Overall, PCA either maintains the accuracy of all the four algorithms or increases the accuracy except for Random Forest at ratios of 7:3 and 6:4.

For the leukemia dataset experiment, we observe from Table 6 that the highest accuracy is achieved for the training to test ratio of 9:1.

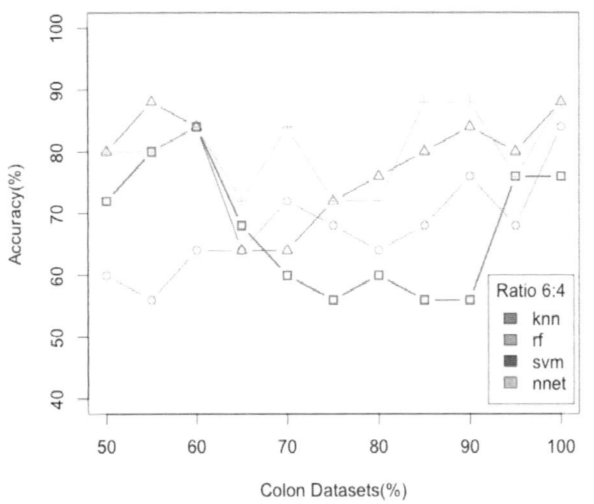

Figure 6. Accuracy comparison in Ratio 6:4

In training to test ratio of 9:1, PCA increased the accuracy of KNN from 28.6% to 100 at the thresholds of 50%, 55% and 70%, an increase of 71.4% in accuracy. PCA increased the accuracy of Random Forest from 14.3% to 100% at threshold of 50%, 55%, 60%, 65%, 70%, 75%, an increase of 85.7% in accuracy. PCA increased the accuracy of Neural Network from 71.4% to 100% at threshold of 50%, 55% and 70%, an increase of 28.6% in accuracy. PCA maintains the accuracy of SVM at 100% at a threshold of 50%, 60%, 65%, 70%, 85%, 90% and 95%.

In training to test ratio of 8:2, PCA increased the accuracy of KNN from 35.7% to 100% at a threshold of 70%, an increase of 64.3% in accuracy. PCA increased the accuracy of Random Forest from 28.6% to 92.9% at threshold of 50%, 55%, 60% and 65%, an increase of 64.3% in accuracy. PCA increased the accuracy of Neural Network from 42.9% to 100% at thresholds of 50%, 55%, 60%, and 65%, an increase of 57% in accuracy. PCA maintains the accuracy of 100% for SVM at a threshold of 95%.

In training to test ratio of 7:3, PCA increased the accuracy of KNN from 54.5% to 95.5% at threshold of 65% and 70%, an increase of 41%. PCA increased the accuracy of Random Forest from 59% to 95.5% at thresholds of 50%, 55%, 60%, 65%, and 70%, an increase of 36.5% in accuracy. PCA increased the accuracy of Neural Network from 63.6% to 100% at thresholds of 50%, 60%, and 65%, an increase of 36.4% in accuracy. PCA maintains the accuracy of SVM at 100% at a threshold of 70%.

In training to test ratio of 6:4, PCA increased the accuracy of KNN from 58.6% to 93% at a threshold of 70%, an increase of 34.4% in accuracy. PCA increased the accuracy of Random Forest from 55% to 96.5% at thresholds of 50% and 65%, an increase of 41.5% in accuracy. PCA increased the accuracy of Neural Network from 68.9% to 100% at thresholds of 50%, 55%, 60% and 65%, an increase of 31% in accuracy. However, the accuracy of SVM decreased from 100% to 96.5% at thresholds of 50%, 55%, 70% and 95%.

From the two datasets that PCA increases the accuracy of the four classifiers at different thresholds. There are significant improvements in the accuracy for leukemia dataset.

6 Conclusions

In this paper, we applied the Principle Component Analysis (PCA) on colon dataset and the leukemia dataset and we compared the accuracy for four different classifiers. Support Vector Machine and Neural Network gave the best performance among the four methods. The experiments included 10-fold cross validation and different training to test ratios of 9:1, 8:2, 7:3 and 6:4.

PCA increased the accuracy of the four classifiers for the colon and leukemia datasets. However, it was observed that there were significant improvements in the performance of most of the classifiers with 10-folds cross validation. The improvements were more significant for the leukemia dataset. In the case of different training to test ratios, PCA maintained the accuracy of the classifiers or increased the accuracy for the colon dataset. However, PCA increased the accuracy of the classifiers significantly for the leukemia dataset.

PCA was selected as a feature selection method to test for increase in accuracy of classifiers on test datasets. The results were promising and it gives us further incentive to test the accuracy of the classifiers with other feature selection algorithms in future.

7 References

[1] A. Richard and Young, "Biomedical Discovery with DNA Arrays," Cell, vol. 102, pp. 9-15, 2000.

[2] Liu, Hsi-Che, Pei-Chen Peng, Tzung-Chien Hsieh, Ting-Chi Yeh, Chih-Jen Lin, Chien-Yu Chen, Jen-Yin Hou, Lee-Yung Shih, and Der-Cherng Liang. "Comparison of Feature Selection Methods for Cross-Laboratory Microarray Analysis." *IEEE/ACM Trans. Comput. Biol. and Bioinf. IEEE/ACM Transactions on Computational Biology and Bioinformatics* 10.3 (2013): 593-604. Web.

[3] Alon, U., N. Barkai, D. A. Notterman, K. Gish, S. Ybarra, D. Mack, and A. J. Levine. "Broad Patterns of Gene Expression Revealed by Clustering Analysis of Tumor and Normal Colon Tissues Probed by Oligonucleotide Arrays." *Proceedings of the National Academy of Sciences* 96.12 (1999): 6745-750. Web.

[4] Golub, T. R. "Molecular Classification of Cancer: Class Discovery and Class Prediction by Gene Expression Monitoring." Science 286.5439 (1999): 531-37. Web.

[5] Dupuy, A., and R. M. Simon. "Critical Review of Published Microarray Studies for Cancer Outcome and Guidelines on Statistical Analysis and Reporting." *JNCI Journal of the National Cancer Institute* 99.2 (2007): 147-57. Web.

[6] Subramanian, J., and R. Simon. "Gene Expression-Based Prognostic Signatures in Lung Cancer: Ready for Clinical Use?" *JNCI Journal of the National Cancer Institute* 102.7 (2010): 464-74. Web.

[7] Sun, Z., D. A. Wigle, and P. Yang. "Non-Overlapping and Non-Cell-Type-Specific Gene Expression Signatures Predict Lung Cancer Survival." *Journal of Clinical Oncology* 26.6 (2008): 877-83. Web.

[8] Pusztai L, Symmans FW, Hortobagyi GN. Development of pharmacogenomic markers to select preoperative chemotherapy for breast cancer. Breast Cancer2005;12:73–85.

[9] Verderio, P. "Assessing the Clinical Relevance of Oncogenic Pathways in Neoadjuvant Breast Cancer." *Journal of Clinical Oncology* 30.16 (2012): 1912-915. Web.

[10] Howley, Tom, Michael G. Madden, Marie-Louise O'Connell, and Alan G. Ryder. "The Effect of Principal Component Analysis on Machine Learning Accuracy with High Dimensional Spectral Data." *Applications and Innovations in Intelligent Systems XIII* (2006): 209-22. Web.

[11] O'connell, Marie-Louise, Tom Howley, Alan G. Ryder, Marc N. Leger, and Michael G. Madden. "Classification of a Target Analyte in Solid Mixtures Using Principal Component Analysis, Support Vector Machines, and Raman Spectroscopy." Opto-Ireland 2005: Optical Sensing and Spectroscopy (2005): n. pag. Web.

[12] Petránek, S. "Neural Network Based Principal Components Analysis for EEG Pre-processing and Analysis." *Electroencephalography and Clinical Neurophysiology* 103.1 (1997): 115. Web.

[13] Raychaudhuri, Soumya, Joshua M. Stuart, and Russ B. Altman. "Principal Components Analysis To Summarize Microarray Experiments: Application To Sporulation Time Series." Biocomputing 2000 (1999): n. pag. Web.

[14] Hilsenbeck, S. G., W. E. Friedrichs, R. Schiff, P. O'connell, R. K. Hansen, C. K. Osborne, and S. A. W. Fuqua. "Statistical Analysis of Array Expression Data as Applied to the Problem of Tamoxifen Resistance." JNCI Journal of the National Cancer Institute 91.5 (1999): 453-59. Web.

[15] Vohradsky, Jiří, Xin-Ming Li, and Charles J. Thompson. "Identification of Procaryotic Developmental Stages by Statistical Analyses of Two-dimensional Gel Patterns." Electrophoresis 18.8 (1997): 1418-428. Web.

[16] Abdi, H. and Williams, L. J. (2010). Principal component analysis. WIREs Comp Stat, 2: 433–459. doi: 10.1002/wics.101

[17] Antoniadis, A., S. Lambert-Lacroix, S., Leblanc, F., Effective dimension reduction methods for tumor classification using gene expression data Bioinformatics (2003) 19 (5): 563-570

doi:10.1093/bioinformatics/btg062

Table 5. Ratio Validation Results For Colon Dataset

		9:1		8:2		7:3		6:4	
	Method	*accuracy*	*auc*	*accuracy*	*auc*	*accuracy*	*auc*	*accuracy*	*auc*
Raw	*Knn*	83.33%	0.75	83.33%	0.75	78.95%	0.683	76%	0.7
	Svm	100%	1	91.667%	1	89.47%	0.9	88%	0.93
	Rf	100%	1	100%	1	89.47%	0.892	84%	0.9
	Nnet	100%	0.85	100%	0.916	89.47%	0.935	88%	0.893
95%	*Knn*	83.33%	0.875	83.3%	0.875	78.95%	0.867	76%	0.775
	Svm	100%	1	91.67%	1	89.47%	0.917	80%	0.91
	Rf	66.7%	0.875	66.67%	0.797	68.42%	0.725	68%	0.77
	Nnet	100%	1	83.3%	0.875	73.68%	0.800	76%	0.81
90%	*Knn*	83.33%	0.875	66.67%	0.688	57.89%	0.642	56%	0.575
	Svm	100%	1	100%	1	89.47%	0.95	84%	0.92
	Rf	66.67%	1	75%	0.734	73.68%	0.833	76%	0.845
	Nnet	100%	1	100%	1	94.74%	0.983	88%	0.91
85%	*Knn*	66.67%	0.625	58.33%	0.563	63.16%	0.675	56%	0.575
	Svm	100%	1	91.67%	1	84.21%	0.95	80%	0.910
	Rf	83.33%	1	83.33%	0.875	63.16%	0.783	68%	0.82
	Nnet	100%	1	91.67%	1	84.21%	0.967	88%	0.88
80%	*Knn*	66.67%	0.5	58.33%	0.438	63.16%	0.4	60%	0.45
	Svm	83.33%	1	91.67%	0.938	84.21%	0.850	76%	0.84
	Rf	83.33%	1	83.33%	0.969	68.42%	0.817	64%	0.83
	Nnet	100%	1	83.33%	0.938	73.68%	0.833	72%	0.77
75%	*Knn*	66.67%	0.5	58.33%	0.438	63.16%	0.4	56%	0.35
	Svm	66.67%	0.75	91.67%	1	78.95%	0.842	72%	0.87
	Rf	83.33%	1	91.67%	1	84.21%	0.833	68%	0.8
	Nnet	100%	1	66.67%	0.906	78.95%	0.817	72%	0.78
70%	*Knn*	66.67%	0.5	58.33%	0.438	63.16%	0.4	60%	0.375
	Svm	100%	1	66.67%	0.875	73.68%	0.875	64%	0.835
	Rf	83.33%	0.875	66.67%	0.906	78.95%	0.85	72%	0.94
	Nnet	66.67%	1	66.67%	0.906	78.95%	0.833	84%	0.92
65%	*Knn*	66.67%	0.5	75%	0.625	73.68%	0.558	68%	0.575
	Svm	100%	1	75%	0.938	78.95%	0.858	64%	0.81
	Rf	83.33%	1	83.33%	0.938	73.68%	0.933	64%	0.835
	Nnet	100%	1	83.33%	1	84.21%	0.917	72%	0.92
60%	*Knn*	83.33%	0.75	83.33%	0.75	84.21%	0.808	84%	0.825
	Svm	100%	1	100%	1	78.95%	0.933	84%	0.95
	Rf	100%	1	83.33%	1	78.95%	0.942	64%	0.955
	Nnet	100%	1	100%	1	84.21%	0.933	84%	0.98
55%	*Knn*	66.67%	0.625	75%	0.688	78.95%	0.775	80%	0.8

	Svm	83.33%	1	91.67%	1	63.16%	0.958	88%	0.95
	Rf	100%	1	91.67%	1	84.21%	0.917	56%	0.96
	Nnet	100%	1	100%	1	78.94%	0.9	80%	0.94
50%	*Knn*	66.67%	0.625	75%	0.688	73.68%	0.65	72%	0.675
	Svm	100%	1	100%	1	89.47%	0.9	80%	0.88
	Rf	100%	1	100%	1	73.68%	0.908	60%	0.95
	Nnet	100%	1	83.33%	1	89.47%	0.883	80%	0.92

Table 6. Ratio Validation Results For Leukemia Dataset

		9:1		**8:2**		**7:3**		**6:4**	
	Method	*accuracy*	*auc*	*accuracy*	*auc*	*accuracy*	*auc*	*accuracy*	*auc*
Raw	*Knn*	28.5714%	0.583	35.7143%	0.550	54.5455%	0.545	58.6207%	0.546
	Svm	100%	1	100%	1	100%	1	100%	1
	Rf	14.2857%	1	28.5714%	1	59.0909%	0.888	55.1724%	0.798
	Nnet	71.4286%	1	42.8571%	0.9	63.6364%	0.983	68.9655%	0.870
95%	*Knn*	71.4286%	0.833	78.5714%	0.775	81.8182%	0.818	72.4138%	0.714
	Svm	100%	1	100%	1	95.4545%	1	96.5517%	1
	Rf	57.1429%	1	35.7143%	0.975	59.0909%	0.921	65.5172%	0.825
	Nnet	71.4286%	1	42.8571%	0.825	72.7273%	0.818	72.4138%	0.731
90%	*Knn*	42.8571%	0.667	50%	0.650	54.5455%	0.545	58.6207%	0.538
	Svm	85.7143%	1	85.7143%	1	90.9091%	0.975	89.6552%	0.976
	Rf	42.8571%	1	50%	1	59.0909%	0.893	65.5172%	0.873
	Nnet	71.4286%	1	64.2857%	0.9	68.1818%	0.851	72.4138%	0.793
85%	*Knn*	42.8571%	0.667	50%	0.650	54.5455%	0.545	62.069%	0.584
	Svm	71.4286%	1	71.4286%	0.975	86.3636%	0.975	89.6552%	0.976
	Rf	85.7143%	1	57.1429%	1	63.6364	1	68.9655%	0.962
	Nnet	57.1429%	1	71.4286%	0.925	72.7273%	0.901	79.3103%	0.870
80%	*Knn*	57.1429%	0.750	57.1429%	0.7	68.1818%	0.682	72.4138%	0.7
	Svm	71.4286%	1	71.4286%	1	90.9091%	0.942	89.6552%	0.962
	Rf	85.7143%	1	71.4286%	1	81.8182%	0.992	72.4138%	0.988
	Nnet	57.1429%	1	85.7143%	1	90.9091%	0.934	72.4138%	0.861
75%	*Knn*	85.7143%	0.917	92.8571%	0.95	81.8182%	0.818	75.8621%	0.752
	Svm	71.4286%	1	85.7143%	0.975	86.3636%	0.934	89.6552%	0.942
	Rf	100%	1	64.2857%	1	77.2727%	0.996	79.3103%	0.981
	Nnet	71.4286%	1	85.7143%	0.975	81.8182%	0.967	79.3103%	0.875
70%	*Knn*	100%	1	100%	1	95.4545%	0.955	93.1034%	0.923
	Svm	100%	1	85.7143%	1	100%	1	96.5517%	1
	Rf	100%	1	78.5714%	1	95.4545%	1	89.6552%	0.955
	Nnet	100%	1	85.7143%	1	90.9091%	1	89.6552%	0.976
65%	*Knn*	85.7143%	0.917	92.8571%	0.95	95.4545%	0.955	89.6552%	0.892
	Svm	85.7143%	1	85.7143%	1	95.4545%	1	96.5517%	1
	Rf	100%	1	92.8571%	1	95.4545%	1	96.5517%	1
	Nnet	85.7143%	1	100%	1	100%	1	100%	1
60%	*Knn*	85.7143%	0.917	85.7143%	0.9	90.9091%	0.909	89.6552%	0.892
	Svm	100%	1	78.5714%	1	90.9091%	1	93.1034%	1
	Rf	100%	1	92.8571%	1	95.4545%	1	89.6552%	1
	Nnet	100%	1	100%	1	100%	1	100%	1
55%	*Knn*	100%	1	71.4286%	0.8	77.2727%	0.773	82.7586%	0.815
	Svm	100%	1	92.8571%	0.975	95.4545%	0.992	96.5517%	0.990
	Rf	100%	1	92.8571%	1	95.4545%	1	93.1034%	1
	Nnet	100%	1	100%	1	95.4545%	1	100%	1
50%	*Knn*	100%	1	78.5714%	0.850	81.8182%	0.818	86.2069%	0.853
	Svm	85.7143%	1	92.8571%	1	95.4545%	1	96.5517%	1
	Rf	100%	1	92.8571%	1	95.4545%	0.996	96.5517%	1
	Nnet	100%	1	100%	1	100%	1	100%	1

SESSION

SIGNAL AND DATA PROCESSING, CLUSTERING METHODS, IMAGING SCIENCE, AND DATA QUALITY ENHANCEMENT

Chair(s)

TBA

Large Scale SVS Images Stitching for Osteosarcoma Identification

B. Armaselu[1*], H.B. Arunachalam[1*], O. Daescu[1], J.P. Bach[2], K. Cederberg[2], S. Glick[2], D. Rakheja[2], A. Sengupta[2], S. Skapek[2] and P. Leavey[2]
*Contact author
[1]Email: {bxa120530, harishb, daescu} @ utdallas.edu
Department of Computer Science, The University of Texas at Dallas, 800 W Campbell Rd, Richardson, TX, USA
[2]Email: {John-Paul.Bach, Kevin.Cederberg, Sam.Glick, Dinesh.Rakheja, Stephen.Skapek}@ UTSouthwestern.edu, Patrick.Leavey@ UTSouthwestern.edu, ANITA.SENGUPTA@childrens.com
University of Texas Southwestern Medical Center, 5323 Harry Hines Blvd, Dallas, TX, USA

Abstract— *We are given a gross map of a bone specimen of a childhood osteosarcoma case, as well as a dataset consisting of digital Whole Slide Images (WSIs) of portions of the bone. The WSIs are in SVS format and they have very high resolutions (between 20x and 40x and up to 65000 x 65000 pixels). We design three applications. The first of them performs digital stitching of these WSIs into one big WSI corresponding to the gross map, based on an image stitching algorithm, which we have developed and will describe in this paper. The second application is a webpage based application that displays the stitching result. Finally, the third application allows the user to navigate and visualize portions of a WSI image or stitching result, at different levels of magnifications. To the best of our knowledge, these are the first results for SVS WSI images stitching.*

Keywords: svs, whole slide image, image stitching, osteosarcoma

1. Introduction

Osteosarcoma is one of the most common types of childhood bone cancer. In order to effectively estimate the likelihood of treatment success, the region of interest of the affected bone is physically cut into pieces which are then scanned and converted to digital Whole Slide Images (WSIs). For proper diagnosis, whole slide images are taken at resolutions between 20x and 64x and it is desirable to stitch them back into one big image of the whole bone specimen. The WSI images are in SVS format and have a very large number of pixels (up to 65000 x 65000). Given a gross map of the whole bone (in JPEG format), which is divided by grid lines into labeled regions, and a set of WSIs in SVS format, one for each region, called *region WSIs*, the goal is to stitch the region WSIs into one big image for the whole bone, called *whole bone WSI*, which is in JPEG format.

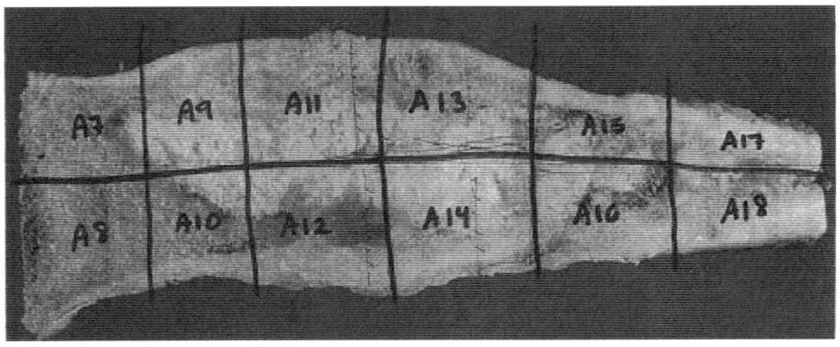

Fig. 1: Global image of the whole bone

1.1 Background and Setup

UT Southwestern Pediatric Oncologists care for 15-20 pediatric patients with osteosarcoma annually at Children's Medical Center in Dallas. For this work, archival samples for 50 patients treated between 1995-2015 were identified. Each sample case has a tumor map (or gross map) and 8 - 50 slides of sections of the bone, which are scanned with Aperio Scanscope© and digitally stored in SVS format [9].

For SVS Image I/O operations, we used the Openslide library, which provides the possiblity to read an SVS image from a file and create a scaled down JPEG image. It does not, however, provide the possibility to process the SVS image directly, convert a JPEG image into an SVS image, or write an SVS image to a file.

Every SVS image has several layers, each corresponding to a level of maginifcation. The *thumbnail level* of the image is the image at 1X magnification. The full resolution image (20X to 64X) is also called *base level image*.

1.2 Related work

The problem of image stitching is an interesting problem and has been studied by various researchers. For the case when images have fixed sizes and orientations, Levin et. al [4] give an image stitching approach based on a window

Fig. 2: Detailed whole slide image of a portion of the bone (A12), in proper orientation

of overlap of width δ and minimizing the difference of the gradient of the images within the window. For arbitrarily sized and fixed orientation images, Rankov et. al [6] designed an image stitching algorithm based on pairwise image cross-correlation of the pixels within a sliding window of overlap. The goal here is to position the window so as to maximize the cross-correlation. For the arbitrary orientation version, Gallagher [2] gave a method for solving puzzles with square pieces. The method involves computing the Mahalanobis Gradient between two images to calculate pairwise compatibility. Kruskal's Minimum Spanning Tree technique is then used to stitch the images, in which each node represents images of specific orientation and each edge represents compatibility scores between nodes. Ma et al [5] developped an application to seamlessly stitch and visualize electron microscopy images (showing cancer or infection). Armaselu et. al [1] develop an algorithm for stitching JPEG images (up to 1000 x 1000 pixels in size) of arbitrary sizes and orientations. The algorithm is based on a quad detection procedure which is run on the gross map, and maximizing neighboring image correlations to find the appropriate orientations of the images. Zerbe et al [8] develop a distributed image processing and analysis framework that admits histological large scale WSI images. Their main application of the framework is immage sharpness assessment.

What is worth mentioning of all these techniques (except the one in [1]) is that the images have either fixed sizes or fixed orientations, which is not the case in our datasets. In addition, the sizes of the images are relatively small (usually up to 500x500 pixels per puzzle piece and up to 5000x5000 pixels overall). In our datasets, WSI images in SVS format take up to 65000x65000 pixels each. If we were to place them together in proper position, the final image would take as much as 500K x 500K pixels overall.

1.3 Challenges

Stitching WSIs to fit a lower resolution gross image for osteosarcoma specimens is a difficult problem due to a number of issues such as:

1. WSI are large themselves (1 - 3 billion pixels), so the final image would be huge (up to 150 billion pixels). Computing it would require up to 450 GB of RAM, which is unachievable for almost any PCs. In practice, we scale WSIs to a smaller level of magnification (such as 1X and 4X instead of 64X), and then do the stitching.
2. Region WSIs need to be rotated, oriented, and scaled consistently with the corresponding properties of the bone region they are taken from.
3. Whole Slide Images in osteosarcoma include extra white spaces around their boundaries and other artifacts, due to the bone cutting process. Since most stitching approaches use the correlation of pixel values within the overlapping window in order to find the best match, if there are white spaces, the correlation would be the same for all images (0), so the best match could not be found. Besides, stitching cannot be done if the margins are white spaces only.
4. The WSI images have high resolutions while the gross image to reconstruct is low resolution. Current approaches do not downsize the images to match the gross image. They also don't use information from the gross image, regarding which WSI belongs to which block.
5. In some instances, WSI Images also need to be re-skewed to obtain proper alignment.

1.4 Our Results

We provide an algorithm to perform image stitching of the thumbnail level images of the WSIs. Stitching is performed based on pairwise WSI pixel and gradient correlation, as well as special markers found in WSIs. We also develop three applications: a WSI Stitching application that implements our stitching algorithm, a webpage-base application to render the stitching result and finally, an image navigation application, which allows visualizing, zooming and and navigating portions of the whole WSI image. We have also tested our image stitching algorithm on the cases of the dataset.

2. Image Stitching

We start by giving an overview of our method (see the System Architecture in Figure 3). Given the global image I of the bone, we first rotate it such that the marked lines, along which the pieces will be cut, are axis-aligned. Then we detect quads defined by lines (and usually labeled by numbers). Quads are sorted from left to right and each quad q. maintains a list of neighboring quads and their span within the border of q. After that, we read all SVS images and take their thumbnail level WSI (at 1X magnification), which are in JPEG format. In the following, we will refer to these thumbnail WSI's as region WSIs. Then, we insert the first region WSI R_0, and for each subsequent region WSI R_i, we consider the quad q_i corresponding to R_i and its left neighbors. Newer dataset images have a blue arrow that points up when the image is properly orientated (see Figure 4). We first try to detect the blue arrow in the

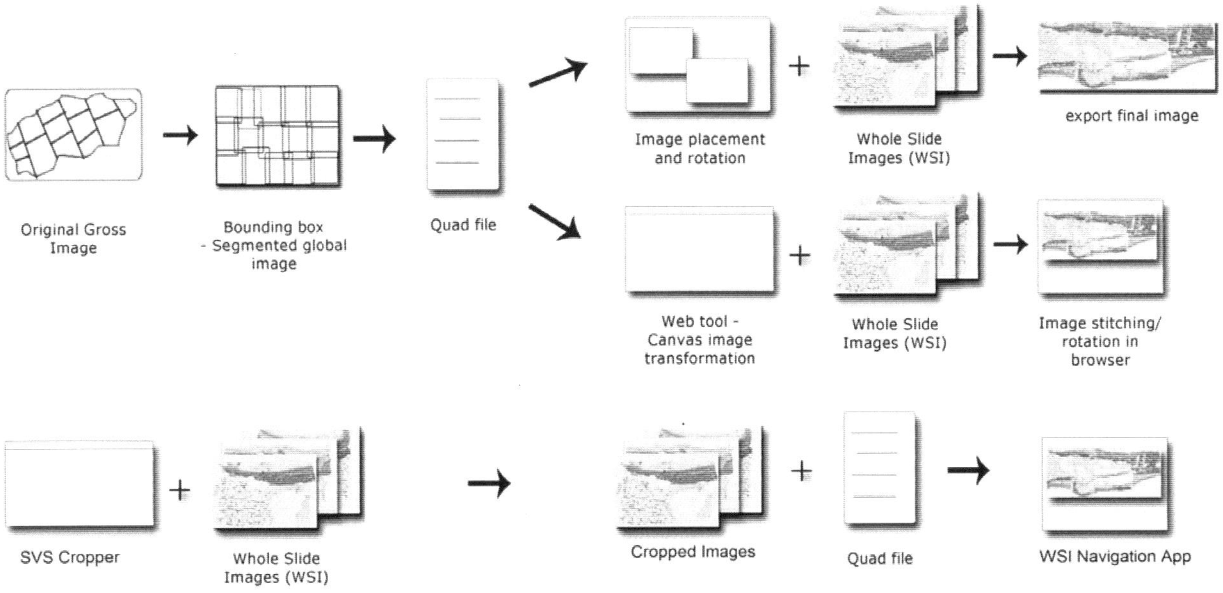

Fig. 3: System Architecture

image, along with its orientation. If we succeed, we rotate the image such that the arrow points up, then restrict the image to the bounding box of the pixels that are non-white and non-blue. Otherwise, for each left neighbor q_j, we take the span (or window) w_{ij} of overlap between q_i, q_j, and we perform image orienting and rotation based on pairwise image correlation between R_i and R_j, as described in [1]. The most common rotation / orientation obtained for each neighbor is selected. Finally, for each pair of images R_i, R_j, with R_j to the left of R_i, we perform seamless image stitching, based on pairwise image gradient matching within the window of overlap [1]. We do this to ensure that the whole bone WSI does not have visible "cuts" (or seams). Although most whole slide images are fine-grained, and thus visible cuts are unlikely, we still do this to make sure it works reasonably well in all cases. The result of the stitching process described is the whole bone WSI and is saved into a JPEG file. Its resolution is smaller than the total number of pixels of all SVS images, but larger tha that of the gross map. Typically, the whole bone WSI has up to 8000 x 8000 million pixels.

The process above is done after the thumbnail image is taken of each SVS image, so the stitching is done at the thumbnail level. Note that, due to space constraints, we cannot afford to stitch directly at high levels of magnification. For the image navigation application, we do, however, stitch the images that are to be displayed within the image

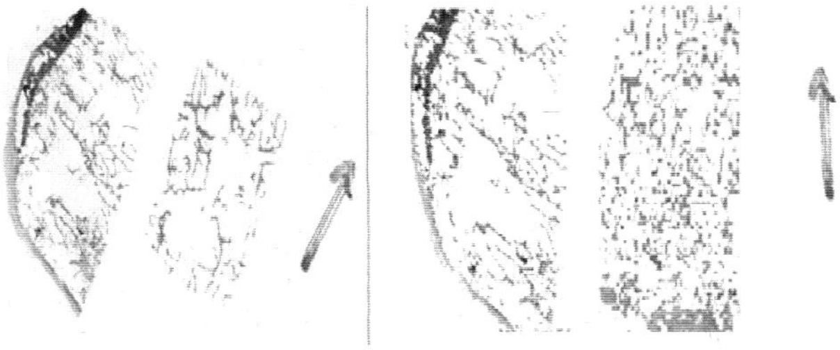

Fig. 4: Thumbnail of a WSI image with blue arrow (left side). The proper orientations are the ones in which the blue arrows point up (right side)

navigation window. This way the user can visualize high-resolution images.

2.1 Quad detection

Given an image, we want to retrieve a list of quads which are bounded by dark lines in the gross image. To do this, we can use the Matlab script in [1] to find the quads using the Hough Line detector. Unfortunately, this approach is not very reliable, because of the following. The global image is blurred and the Hough Transform would return a lot of irrelevant lines, as well as omit a few relevant ones. To find a good threshold (minimizing number of irrelevant lines, as well as omitted relevant lines) is hard (sometimes impossible) and depends very much on the dataset. To avoid these issues, we have a better approach.

We say that a pixel is *dark* if the V coordinate of the HSV color of the pixel is less than a certain threshold L (in practice we got the best results for $L = 32$ out of 256).

We use a **merging algorithm** as follows. Each pixel (i, j) is assigned an integer label $l(i, j)$. At each step, if neighboring pixels with distinct positive labels l_1 and l_2 ($l_1 < l_2$) are found, then their quads are merged, i.e. for all pixels (i, j) for which $l(i, j) = l_2$, we assign $l(i, j) = l_1$. This is done in 3 steps:

1) Let $l = 1$. Traverse the image in row order and assign label l to non-dark pixels. Once a dark pixel is encountered, the label l is incremented and the next dark pixels are skipped until a non-dark pixel is encountered.

2) Traverse image in column order and, for every adjacent non-dark pixels p, q, assign $\min\{l(p), l(q)\}$ to both p, q.

3) Create regions and assign pixels to them based on their label (pixel p is assigned to region k if $l(p) = k$).

4) Discard irrelevant regions (which are smaller than an area threshold A. In practice we used $A = w \cdot h/1250$, where w, h are the width and height of the image). Rather than counting pixels in a region, we take the area of the bounding box of the region, which is discarded if the area is less than A.

Finally, after quads have been detected, we compute, for each quad, the list of neighboring quads and we sort all quads in left-to-right order.

2.2 Blue arrow detection

Given an image, the goal is to find a blue arrow in the image.

We first restrict the image to the axis-aligned bounding box B of the blue pixels (where "blue" is defined by a threshold of H, S, V values). Then, the direction of the arrow is detected as follows.

Case 0. If B has fewer than 100 pixels, then there is no arrow

Case 1. If height of B is at least 4 times the width of B, then the arrow is vertical

Case 2. If width of B is at least 4 times the hight of B, then the arrow is horizontal

Case 3. Otherwise, the arrow is diagonal.

To detect the orientation of the arrow, we split B into two halves (4 quadrants if the arrow is diagonal). The half (quadrant) containing the most blue points indicates the orientation of the arrow (e.g. if more blue pixels are in the upper half, then the arrow is pointing up).

If no blue arrow pointing up is present on the image, we rotate it as described in [1] under the Image rotation subsection.

If there is a blue arrow, we rotate it such that the arrow points up, then re-skew the image such that its non-white edge is as close to horizontal as possible.

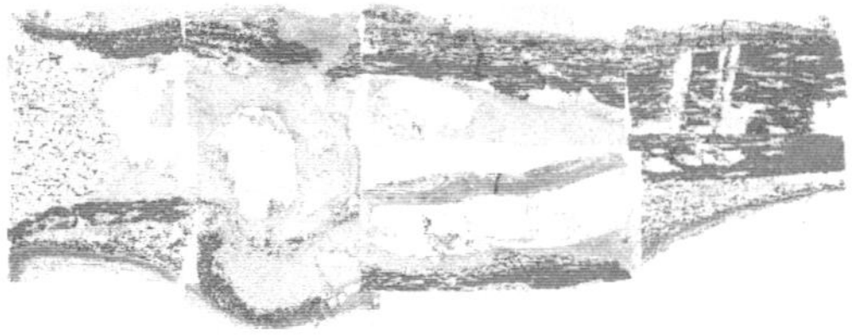

Fig. 5: This image shows the whole bone WSI of case 6 displayed in the web-based visualization application

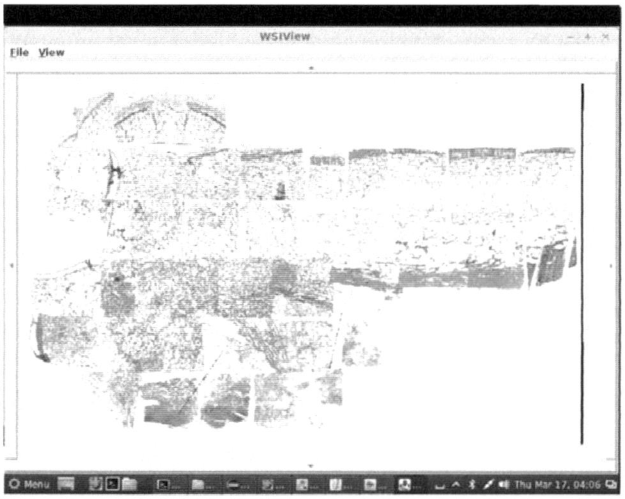

Fig. 6: The image navigation application showing the thumbnail level stitching result

Let θ be the angle by which the image is rotated. θ is computed with an error of at most $error_\theta = \text{atan}(\frac{w}{2l})$, where l is the length of the blue arrow and w is its width. In most cases, $l/w > 10$, so $error_\theta < 3^o$.

3. Image Navigation and Visualization

We develop an application to allow navigation of the whole bone WSI. The possible actions the user can take are:

1) move up, down, left, right one tile

2) zoom in / out

3) view the whole image at different levels of magnification (as long as JVM has enough memory to allow this).

The levels of magnification are 1X, 8X. For 8X, the image is divided into tiles, and a minimap of the whole image is displayed on the right side (Figure 6). The current view is indicated as a black window on the minimap. Typically, the tiles are 64x64 pixels, and the minimap is at 1:16 scale. For 1X (which is the default level of magnification), the program displays the thumbnail level stitching result, with no minimap (See figure 5).

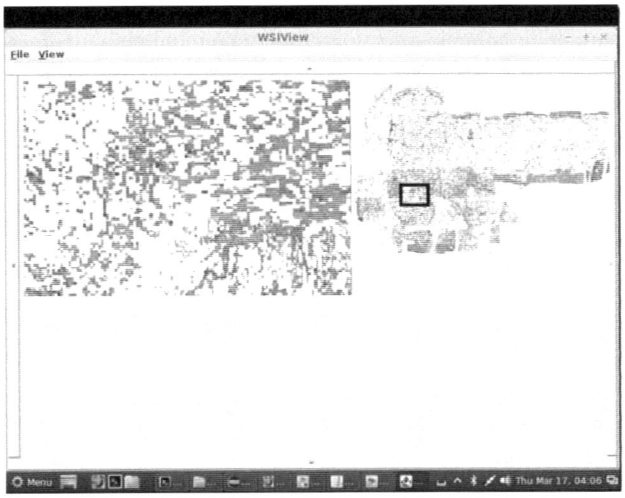

Fig. 7: The image navigation application with WSI magnified at 8X. On the minimap, the black window shows the location of the current view within the WSI

Each time the user takes an action, we consider the coordinates of the window to be displayed and redo the image stitching with the images crossing this window only, then select the tiles within the respective window. To this end, for each initial WSI image, we create a 16 x 16 tiling for each level of magnification, using an SVS cropping software we have also developed (see Figure 3). Once an image is cropped, each tile is saved to a different file. To save space, we also create a mapping M that maps an image coordinate to a pointer to a WSI image in the list of WSI files, eliminating the need to store all images in memory (unless they are used for stitching). Each time the window is updated, we update only the list of WSI images to be stitched.

There is also an issue about rotating WSI images at higher magnification. As discussed earlier, the Openslide API does not provide support for rotating an SVS image. We need to first uncompress it to JPEG. However, this uncompression step uses a lot of memory, which may be too much for the JVM. That is why we resort to the following scheme. We first create a thumbnail of the SVS at 1X magnification and find the arrow orientation angle θ. Back to the SVS, we create a map of 16x16 JPEG tiles for the image and compute, for each of them, the coordinates of the SVS image such that, after rotating the tile images by θ and stitching them back, they would roughly correspond to the rotated 8X image.

After performing the quad detection on the gross map, we save them to a file. Each quad record $Q(id, x, y, w, h)$ contains 5 numbers: quad id, top left corner X and Y coordinates (x, y), width and height (w, h). Each WSI image maps to a specific block in the gross image. We then use the web-based HTML/JS application presented in [1] to

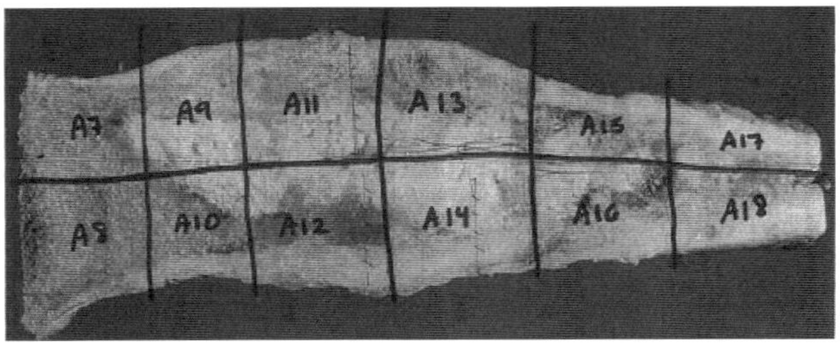

Fig. 8: Gross image of the whole bone - Case 1. The bone is cut into slices numbered A7 through A18, in column-row order

Fig. 9: Output of the image stitching algorithm, when run on Case 1. There is a minor artifact in the top-right portion of the image, which is reflected in the whole bone WSI

render the WSIs in the correct position. The HTML/JS application allows the user to flip and rotate WSIs in 90 degree increments [1]. Figure 7 shows how the whole bone WSI is displayed in the HTML application.

4. Experimental Results

We tested our applications on a given dataset and the results validated our approach. Each case in the dataset contains a JPEG file with the map (or gross image) of the whole bone and a list of m slide images in SVS format. For every such case, we have run all our applications, and we show the results. Due to space limitations, we hereby show only 5 cases (1 through 5, see Figures 8 - 17). We also

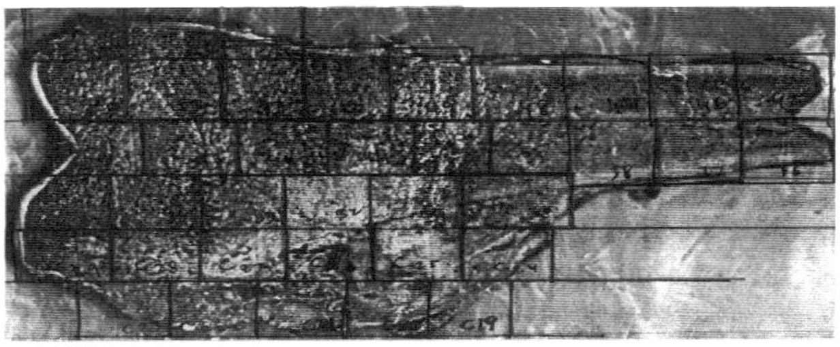

Fig. 10: Gross image of the whole bone - Case 2. The bone is divided into regions numbered C19 through C56, in reverse row-coumn order (starting from bottom right)

Fig. 11: Output of the image stitching algorithm, when run on Case 2. The output image matches the gross image, with little extra white space

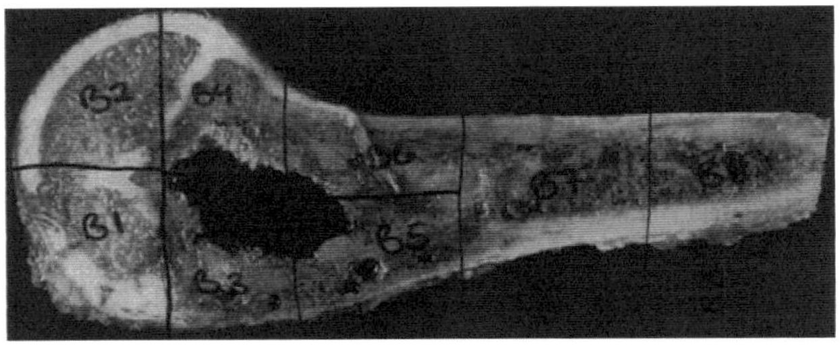

Fig. 12: Gross image of the whole bone - Case 3. The bone is sliced into regions labeled B1 through B8, starting from bottom left

show examples of our image navigation application and the web-based application.

We argue that in the image resulted from the stitching, the portions corresponding to regions have reasonably little extra white spaces (that are not part of the correct final image). Indeed, since extra white spaces are due to WSI rotation only, we guarantee that placing a region WSI R_i within the whole bone WSI leaves a strip of white space of a width at most $\frac{1}{2} \cdot \tan(error_\theta) \cdot width(R_i)$. In most cases, this is less than $\frac{1}{40} \cdot width(R_i)$. However, the border may not seem to be smooth (can "jump" a few pixels).

Our approach performs better than PIDB [7] in terms of memory footprint. PIDB requires a lot of pre-processing and stores processed intermediate images as TIFF files. Our method doesn't generate new TIFF files and hence has low memory footprint.

Table 1: The properties of different case

Dataset No.	Region count	Gross image size	Total WSI size
1	12	8.88E+004	3.69E+010
2	38	1.95E+005	7.59E+010
3	8	3.76E+005	2.00E+010
4	28	7.92E+004	7.40E+010
5	24	1.02E+005	6.35E+010

Table 1 shows the properties of different case. For each

Fig. 13: Output of the image stitching algorithm, when run on Case 3. There is a significant difference between the gross image and the stitched image. The white space in the center of image is due to the presence of large number of artifacts like missing bone portions from slides under study. This makes the problem of image stitching more harder as the correlations do not match correctly between adjacent images.

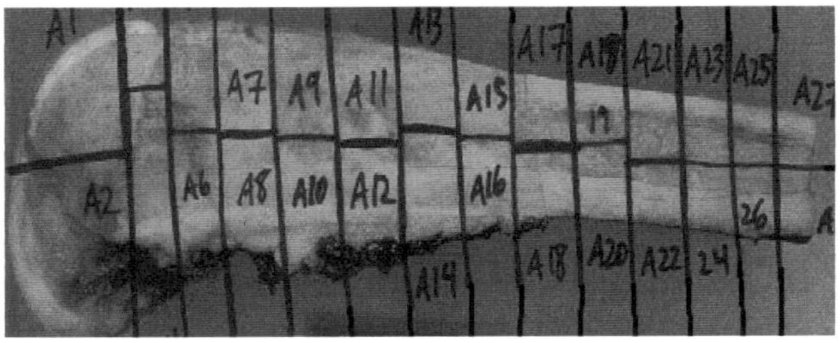

Fig. 14: Gross image of the whole bone - Case 4. The regions of the bone, numbered A1 through A28, are column-row order

case, the table shows the number of regions, the gross image size (in pixels) and the total number of pixels in all the SVS WSIs.

Table 2 shows the performance of the stitching program on different case. For each case, the table shows the running time required for SVS input, for the image stitching itself (in seconds) and the accuracy of the stitching, measured by the percentage of extra white spaces present in the whole bone WSI.

Table 2: The performance of the stitching algorithm when run on different case

Dataset No.	SVS input time(s)	Stitching time (s)	Stitching time per WSI (s)	Extra white space
1	33.30	14.50	1.21	0.83%
2	95.51	8.70	0.23	1.12%
3	11.68	9.90	1.24	0.94%
4	24.48	8.14	0.29	0.71%
5	44.60	7.30	0.30	2.50%

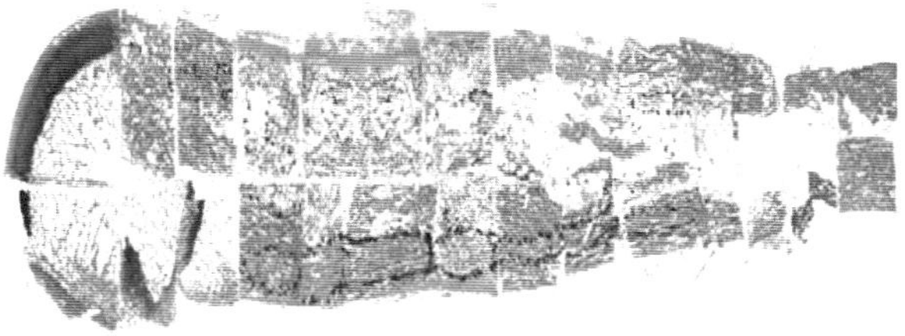

Fig. 15: Output of the image stitching algorithm, when run on Case 4. The final image matches the gross image well, with minor artifacts, which were present in the corresponding input WSIs

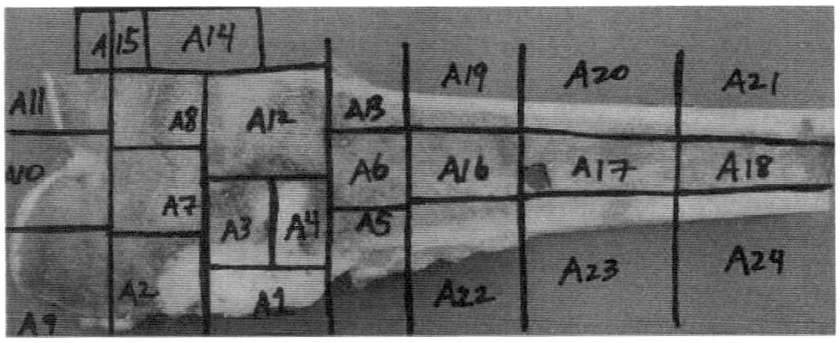

Fig. 16: Gross image of the whole bone - Case 5. The regions of the bone are labeled A1 through A24

5. Conclusion

We have designed an algorithm to digitally stitch a set of WSIs to match a given gross map of the bone specimen. We also develop an application that implements our algorithm and saves the stitching result to a file, an application that renders it on a web page, and an application that allows navigation and visualization of a portion of a WSI or the whole bone WSI. The applications prove to be very useful in practice. This is an important step especially in working with proprietary image formats and we believe the methods will open up a myriad of analysis capabilities for researchers to collaborate. The techniques discussed in this paper can serve as a ground zero in moving away from proprietary software dependencies for image analysis. We plan to couple this method of SVS image stitching with image processing techniques and hope to expand the support to include more proprietary medical image formats. This will help in removing the biggest barrier that every medical researcher faces when working with non-standard medical imaging formats, eventually leading to better problem solving in the field of pathology informatics.

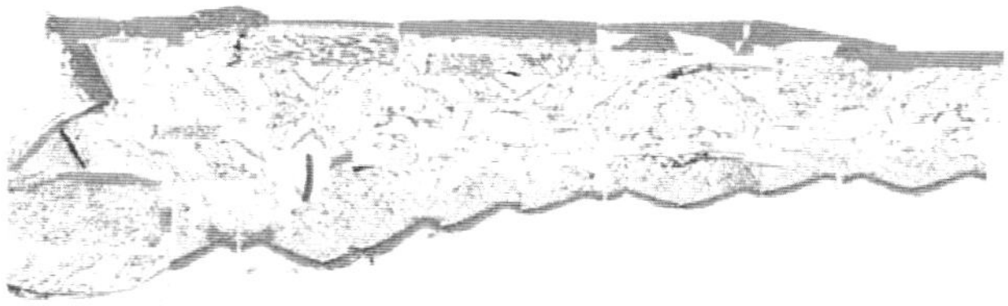

Fig. 17: Output of the image stitching algorithm, when run on Case 5

Acknowledgment

This research was partially supported by NSF award IIP1439718 and CPRIT award RP150164.

References

[1] B. Armaselu, H.B. Arunachalam, O.Daescu, J.P. Bach, K. Cederberg, D. Rakheja, A. Sengupta, S. Shapek, P. Leavey. *WSI Images Stitching for Osteosarcoma Detection*. International Conference on Computational Advances in Bio and Medical Sciences, 2015.

[2] Andrew C.Gallagher. *Jigsaw puzzles with pieces of unknown orientation*. Computer Vision and Pattern Recognition (CVPR), 2012 IEEE Conference on. IEEE, 2012.

[3] Farzad Ghaznavi, Andrew Evans, Anant Madabhushi, Michael Feldman. *Digital imaging in pathology: whole-slide imaging and beyond*. Annual Review of Pathology: Mechanisms of Disease 8 (2013): 331-359.

[4] A. Levin, A. Zomet, S. Peleg, Y. Weiss. *Seamless Image Stitching in the Gradient Domain*. In Proc. ECCV, May 2004

[5] Ma, Bin, Timo Zimmermann, Manfred Rohde, Simon Winkelbach, Feng He, Werner Lindenmaier, and Kurt EJ Dittmar. *Use of autostitch for automatic stitching of microscope images*. Micron 38, no. 5 (2007): 492-499.

[6] V. Rankov, R.J. Locke, R.J. Edens, P.R. Barber, B. Vojnovic. *An algorithm for image stitching and blending*. In Proc. SPIE - Vol. 5701, March 2005.

[7] Wang, F., Oh, T. W., Vergara-Niedermayr, C., Kurc, T., Saltz, J. *Managing and Querying Whole Slide Images*. In Proceedings of SPIE (2012), 8319, 83190J.

[8] Zerbe, Norman, P. Hufnagl, and K. Schluns. *Distributed computing in image analysis using open source frameworks and application to image sharpness assessment of histological whole slide images*. Journal of Diagnostic Pathology, 2011, Vol. 6 (Suppl. 1): S1-S26

[9] http://www.openmicroscopy.org/site/support/bio-formats5.1/formats/aperio-svs-tiff.html

Optimizing the Detection of Events of Interest in Serial Data

Robert A. Warner, MD

Tigard Research Institute
12228 SW Chandler Drive
Tigard, OR 97224-2825
USA
hillwarner@frontier.com
Contact Author: Robert A. Warner, MD

1.0 Abstract

This paper describes two methods for improving the detection of events of interest in serially acquired data: 1. combining the use of Z scores with the central moving average (CMA) and 2. the iterative use of the CMA. The methods were tested using ECG data to discriminate patients with prior inferior myocardial infarction (IMI) from normal patients and from patients with prior anterior myocardial infarction. The study shows that the combination of the Z score and CMA methods significantly improves the accuracy with which the IMI patients are identified. Also, the iterative use of the CMA's facilitates the visual identification of the IMI patients in analog displays of the data.

Topical Key Words: central moving averages, Z scores

2.0 Introduction

It is often important to review and analyze series of data to identify possible occurrences of events of interest. Examples of this include:

1. Detecting episodes of ischemia or arrhythmia in electrocardiographic (ECG) monitoring data recorded from hospitalized or ambulatory patients.
2. Examining seismographic records for patterns that are compatible with impending earthquake or volcanic activity in the near future.
3. Identifying intermittent aberrations in the measured performances of mechanical devices that suggest that the devices might fail at critical times.

Events such as the above are typically associated with values of relevant parameters that are abnormally high or low compared to the baseline values of those parameters. For example, episodes of myocardial ischemia are often identified by the occurrence of greater ST segment displacement of the continuously recorded ECG than is the case when ischemia is not present. Similarly, seismographic waves recorded during as well as shortly before and after earthquakes typically exhibit increased amplitudes and frequencies compared to seismically quiescent periods.

Events of interest in serial data have a finite duration, i.e. are extended in time, and typically have an identifiable beginning and end. Therefore, the abnormally high or low ranges of values of the parameters that characterize the events of interest are clustered during each occurrence of the event. Reviewers of the data seek to accurately identify such clusters of abnormal values and thereby identify each occurrence of the event of interest. Since the sets of data being reviewed are often very large, the methods used to examine the data should be efficient as well as comprehensive and accurate.

Previous work from this laboratory has demonstrated the effectiveness of each of two independent methods of displaying and analyzing data – the use of Z scores and the calculation of the central moving average (CMA) of the raw measurement data.[1-4] In the present study, I have extended this work by testing the hypothesis that the following methods can improve the detection of events of interest in serial data:

1. The combined use of the Z score and the CMA methods in the same set of data
2. The iterative use of the CMA

1.0 Methods

1.1 Description of the Methods

One of the methods used in this study is that of expressing data not as the raw measured values of a parameter being evaluated, but rather as the data's Z scores (also called standard scores). A Z score expresses the value of a data point as the difference between that data point and the mean of a comparison population of the same type of data. The unit that is used to express this difference is the standard deviation of the comparison population. If X is an individual data point and SD is the standard deviation:

Z score of X = (X – mean) / SD (1)

Since the algebraic sign of a standard deviation is always positive, the algebraic sign of a calculated Z score is determined exclusively by whether the numerical value of individual data point X is greater or less than that of the comparison population's mean.

Regardless of its algebraic sign, the absolute value of a Z score is associated with a particular P value (alpha). This means that the Z score of a data point can be used to test the null hypothesis between that data point and the mean of a comparison population. This is analogous to using the Student T test to test the null hypothesis for two populations of data. For example, absolute values of Z scores of $\geq$1.65, $\geq$2.33 and $\geq$3.08 are associated with P values of 0.05, 0.01, and 0.001, respectively.

The method of displaying and analyzing data that is combined with Z scores in this study is the use of the CMA. The CMA is the arithmetic mean of a given number of data points that symmetrically surround the data point for which the CMA is calculated. For example, in a set of sequentially acquired numerical data, the 10-sample CMA of data point #6 in the series is the arithmetic mean of data points 1 through 5 and data points 7 through 11. A previous study showed that when the 10-sample CMA was calculated for a series of numerical data, the accuracy of detecting events of interest in that series was greater than if the raw data had been used. [3,4] With respect to the use of the CMA, the present study differs from the previous one in two important ways. First, rather than calculating the 10-sample CMA using only the raw measured data, I also calculated the CMA using the Z scores of the measured data. Second, I evaluated the iterative use of the 10-sample CMA as follows: a 1st iteration 10-sample CMA is the 10-sample CMA of the raw or Z score data in the series; a 2nd iteration 10-sample CMA is the 10-sample CMA of the previously calculated 1st iteration CMA; a 3rd iteration 10-sample CMA is the 10-sample CMA of the previously calculated 2nd iteration CMA, etc. These iterative calculations of the CMA were performed up to the 5th iteration 10-sample CMA.

1.2 Description of the Data

The data used in the present study consist of computerized ECG measurements obtained from a total of 1080 patients who had undergone cardiac catheterization with coronary angiography at a university medical center in North Carolina. The patients were divided into three subgroups:

- Normal - 464 patients with no evidence of prior myocardial infarction as shown by cardiac catheterization
- IMI - 341 patients with prior inferior myocardial infarction as shown by cardiac catheterization
- AMI – 275 patients with prior anterior myocardial infarction as shown by cardiac catheterization

The patients were grouped with respect to each of these three diagnostic categories. The phenomenon of interest in this study is the prior occurrence of IMI as distinguished from both the Normal subgroup and the AMI subgroup. The ECG parameter used for each patient is the arithmetic mean of the amplitudes (in microvolts) of the voltages recorded in standard ECG Lead aVF during the first 40 ms. of the QRS complex.[5-7] Since the ECG data had been stored at a frequency of 250 Hertz, each 40 ms. interval of the QRS for each patient represents 10 data points. Diminished initial QRS voltage in Lead aVF is known to constitute ECG evidence of prior inferior MI. Conversely, neither normal patients nor patients with prior anterior MI characteristically exhibit diminished voltage during the initial portion of the QRS complex of the Lead aVF.[7]

1.3 Analysis of the Data

To compare the abilities of the raw data and the CMAs to augment the visual discrimination of the IMI subgroup from the Normal and the AMI subgroups, line graphs of the raw data, the Z scores of the raw data and all the CMAs were plotted. For calculating the Z scores for each patient in all three subgroups, the mean and the standard deviation of the data from the 464 patients in the Normal subgroup was used.

In addition, receiver operating characteristic curves were used to determine the respective sensitivities at 100% specificity for prior inferior MI of the raw data and of each of the iterations of CMA. The statistical significance of any apparent difference in diagnostic performance (compared to the raw data) was tested using chi square analysis. To avoid Type 1 errors associated with multiple comparisons in the chi square analysis, a P value < 0.01 was chosen a priori to indicate statistical significance.

2. Results

In the line graphs shown in Panels A through H of Figure 1, the ECG data from the entire Normal subgroup are followed by the data from the entire IMI subgroup and these are followed by the data of the entire AMI subgroup. Panel A shows the raw data, Panel B shows the Z scores of the raw data, Panel C shows the 1st iteration CMA of the raw data, Panel D shows the 1st iteration CMA of the Z scores of the data and Panels E through H show, respectively, the 2nd, 3rd, 4th and 5th iterations of the CMAs of the Z scores of the data.

As expected, all the panels in Figure 1 show that the data in the IMI subgroup tend to have lower values than the data in either the Normal or the AMI subgroup. Comparing the graph of the raw data in Panel A with that of the Z scores in Panel B reveals that both line graphs have identical contours. This is expected because Panels A and B differ only in that they express the same information in different units, i.e. microvolts vs. standard deviations, respectively. The graphs of the 1st iteration CMAs in Panels C and D have smoother contours than the graphs in Panels A and B. This increased smoothness is because the sequential averaging of ten samples of raw data in the CMA moderates the effects on the graphs' contours of unusually high or low values of individual data points. The increased smoothness of the graphs in Panels C and D permits clearer visual discrimination of the IMI data from the normal and the AMI data compared to that in the graphs in Panels A and B. Additional iterations of the use of CMAs as shown in Panels E, F, G and H of Figure 1 show further smoothing of the line graphs.

For identifying events of interest, Figure 1 also demonstrates a major advantage of graphing the Z scores, rather than the raw voltage data. Panels B, D, E, F, G and H each have a horizontal dashed line in the lower portion of the display. This horizontal line represents a Z score of -1.65. As noted above, data exhibiting Z scores at or below this value are statistically significantly lower than the data from normal patients at a P value <0.05. Thus, the graphing of the Z scores rather than the raw data enables one to determine that the data from the IMI patients not only tend to be lower than those of the other subgroups, but tend to be *statistically significantly* lower. Panels D through H of Figure 1 show that by combining the CMA method with the Z score method, the statistical significance of the difference between the IMI and the other data becomes especially apparent. In Panels D through H, all the IMI data are in close proximity to the horizontal dashed line that designates statistical significance. Conversely, all the Normal and AMI data are well above this line.

Table 1 lists the diagnostic performances for detecting prior IMI of each of the raw voltage and Z score parameters that were evaluated. In Table 1, the first column lists the parameters, the second column shows the diagnostic performance exhibited by that parameter as its % diagnostic sensitivity at 100% diagnostic specificity, the third column shows the threshold value required to attain that diagnostic performance, the fourth column reveals the chi square resulting from comparing each of the last seven parameters to the raw voltages in the first row of data and the fifth column indicates the P value associated with each of these chi squares. As expected, the first two rows of data show that the performances of the recorded voltage data and the Z scores of the voltage data are identical. Also as expected, the third and fourth rows of data in Table 1 reveal that the same is true for the 1st Iteration CMAs of the voltage data and of the Z scores of the voltage data. However, data rows 3 through 8 of Table 1 show that the diagnostic performances of all the CMAs are highly statistically significantly superior to those of both the raw voltage data and to the Z scores of the raw voltage data. In keeping with this, Table 1 reveals that the threshold values needed to achieve 100% diagnostic specificity are much lower for the CMAs than they are for the voltage data and for the Z scores of the voltage data.

Figure 1

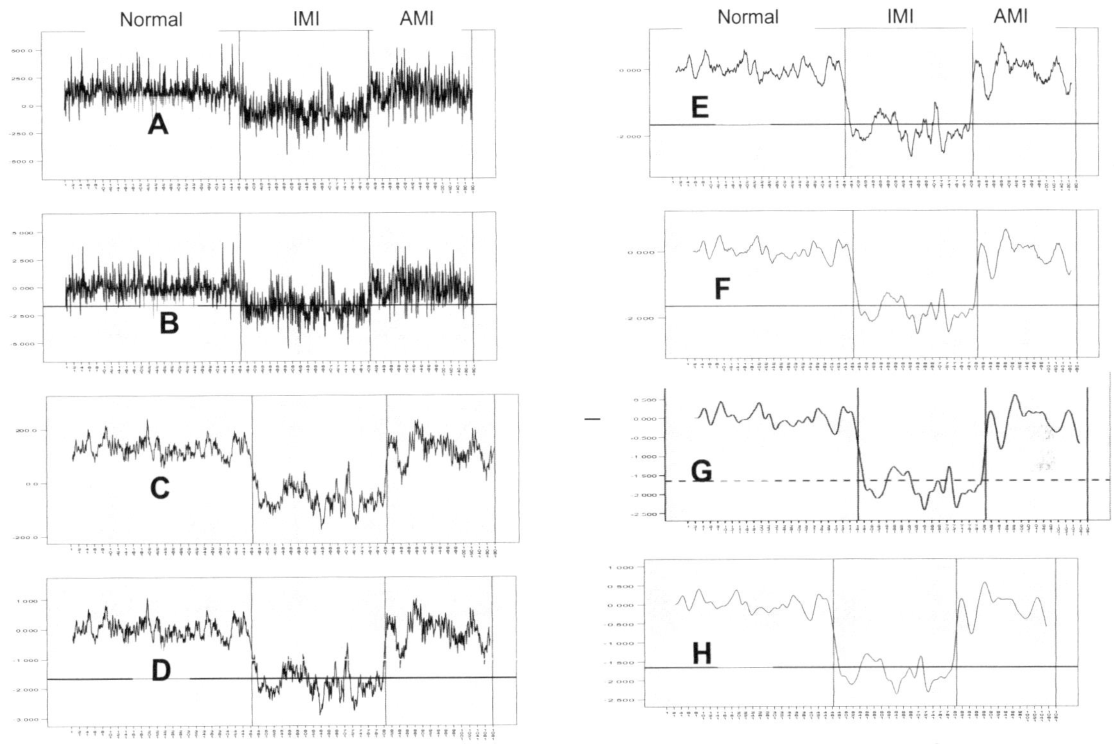

Normal = patients with neither IMI nor AMI, IMI = patients with IMI only, AMI = patients with AMI only. Panel A = raw voltage data, Panel B = Z scores of the raw voltage data, Panel C = 1st Iteration CMA of the raw voltage data, Panel D through H= the 1st through the 5th Iteration CMAs of the Z scores, respectively

Table 1

Parameter	%Sensitivity @ 100% Specificity	Threshold Value	Chi Square*	P*
μV	26.7	-137.4		
Z_ μV	26.7	-2.54	0	NS
1st Iteration CMA of μV	99.1	57.1	383	2.77^{-85}
1st Iteration CMA of Z_μV	99.1	-0.69	383	2.77^{-85}
2nd Iteration CMA of Z_μV	100.0	-0.56	395	6.75^{-88}
3rd Iteration CMA of Z_μV	100.0	-0.56	395	6.75^{-88}
4th Iteration CMA of Z_μV	100.0	-0.63	395	6.75^{-88}
5th Iteration CMA of Z_μV	100.0	-0.63	395	6.75^{-88}

**Compared to the raw voltage in μV. CMA = 10 sample central moving average, μV = microvolts (mean of first 40 ms. of QRS in Lead aVF), NS = not significant, Z_ μV = Z scores of the measured raw voltages*

3. Discussion

It is often important to detect specific events of interest in sequentially acquired data. For example, a very common reason for recording a temporal series of ECG data is the detection of intermittent episodes of cardiac arrhythmia or myocardial ischemia. The detection of events of this type has two major components:

- Visually or algorithmically identifying clusters of data that are consistent with instances of the event of interest and
- Determining that the clusters of data that have been identified in this way represent genuine events rather than artifacts

The present study shows that the combined use of moving averages and Z scores attains both of these goals. Panels C through H of Figure 1 demonstrate that the use of the 10-sample CMA provides much clearer visual separation of the prior IMI data from both the normal and the prior AMI data than does use of the raw data in Panel A or the Z scores of the raw data in Panel B. Panel B and especially Panels D through H of Figure 1 demonstrate an important advantage of plotting the Z scores rather than the raw voltages. Since each absolute value of a Z score is associated with a P value, one can determine if the data that are relevant to an event of interest differ statistically significantly from the baseline data. In the Z score plots in Figure 1, the values that represent the IMI patients closely surround the horizontal dashed line. This means that the IMI data tend to be statistically significantly lower than the data from both the Normal and the AMI subgroups. Figure 1 shows that combining the CMA and the Z score methods of display and analysis is particularly useful. The Z score plot of the raw data in Panel B reveals that many of the data in all three subgroups overlap this line of statistical significance at the $P<0.05$ level of confidence. In contrast, the plots of the CMAs of the Z scores in Panels D through H of Figure 1 much more clearly distinguish the IMI data from the data of the other two subgroups. In Panels D through H, there are no normal or AMI data at or below the horizontal line of statistical significance. However, all the IMI data in Panels D through H are in close proximity to the line of statistical significance.

Panels C through H of Figure 1 also demonstrate the effects of the iterative use of CMAs. As the number of iterations increases from one to five, the smoothness of the graphs of the plotted data also increases. The progressively greater smoothness of the graphs makes the separation of the IMI data from the Normal and AMI data especially obvious. By more clearly separating the data that represent an event of interest from the baseline data, the use of CMAs can facilitate the visual review of serial data. This clearer visual identification of the data that are associated with an event of interest suggests that the iterative use of the CMAs can also facilitate algorithmic identification of similar events of interest. This is because the more complete separation of the data of an event of interest from the baseline data increases the likelihood that highly reliable rules for detecting these events can be developed if CMAs rather than raw data are used. The data shown in Table 1 strongly support such a conjecture. For example, an algorithm developer might wish to establish a diagnostic rule for detecting the IMI patients that exhibits 100% specificity, i.e. produces no false positives. As data row 2 of Table 1 shows, the threshold value of the Z scores of the raw data that produces that result is –2.54. Using that threshold value in an algorithm would detect only 26.7% of the cases of IMI, i.e. produce 73.3% false negatives. In contrast, data row 5 of Table 1 shows that by choosing a threshold value of –0.56 of the 2nd iteration of CMAs would produce no false negatives and no false positives.

Table 1 also reveals that the CMAs of both the raw voltage and the Z score data (the last six rows of data) exhibit a nearly four-fold increase in sensitivity at 100% specificity over the non-averaged data (the first two rows of data). This improvement in diagnostic performance that is produced by using the CMAs is very highly statistically significant.

Previous work from this laboratory showed that for CMAs to be effective in improving the identification of events of interest, the data relevant to the detection of those events must be clustered in groups.[3,8] In the present study, the data are clustered with respect to diagnosis (IMI vs. Normal or AMI). An extremely common example of the clustering of relevant data is the occurrence of identifiable events in time series of data. This is because events that take place in the world are extended in time. Therefore, data that exhibit the ranges of values that are associated with such an event are clustered during each time period during which the event occurs. For example, a criterion for detecting a transient episode of myocardial ischemia in medical monitoring data is the appearance and subsequent disappearance of ST segment displacement in a patient's ECG Throughout the period of ischemia, the abnormal values of ST segment displacement would be temporally clustered. Conversely, the baseline values of the ST segment measurements would be temporally clustered during the non-ischemic periods.

Comparing Panels A and B to Panels C through H in Figure 1 shows an additional advantage of using CMAs rather than raw data. CMAs, especially when used iteratively, permit one to determine very accurately the onset and offset of the event. They also clearly show any changes in the amplitude of the line that represents the values of the relevant parameter during the event. Such observations about the time courses of events of interest can themselves be diagnostically important. This is exemplified by patterns of ST segment displacement of ECG signals that are recorded during episodes of myocardial ischemia. ST segment displacement associated with myocardial ischemia typically persists for at least several minutes, gradually increases in severity, reaches a peak or nadir and then gradually decreases. Conversely, an episode of ST segment displacement that lasts only a few seconds and abruptly reaches its maximum degree of severity is much more likely to represent an artifact, rather an actual ischemic event. Therefore, observing the temporal patterns of ST segment changes in recorded ECG data can provide additional useful diagnostic information by enabling one to apply the principles of conditional probability to the detection of ischemia. If an episode of ST segment displacement has a time course that is physiologically consistent with an ischemic event, it increases the prior probability that the observed ST displacement represents an actual ischemic event rather than an artifact Previous work from this laboratory has shown that applying the principles of conditional probability to medical diagnosis significantly increases the accuracy of diagnostic tests.[9,10]

4. Conclusions

The present study confirms the hypotheses that in a series of ECG data clustered by diagnostic category:

- The combination of Z scores with CMAs significantly improves the ability of ECG data to distinguish prior IMI patients from both normal and prior AMI patients.

- The iterative use of CMAs of relevant data further increases one's ability to discriminate events of interest from surrounding baseline data.
- The iterative use of CMAs also improves the delineation of the temporal features of events of interest in sequentially acquired data. This additional information may further improve diagnostic accuracy by helping to assess the prior probability that an actual event of interest, rather than an artifact, has occurred.

5. References

1. Warner RA. Optimizing the Display and Interpretation of Data. Elsevier, Amsterdam, 2015:7-72
2. Warner RA, Olicker AL, Haisty WK, Hill NE, Selvester RH Wagner GS. The importance of accounting for the variability of electrocardiographic data among diagnostically similar patients. Amer. J. Cardiol. 86:1238-1240, 2000
3. Warner RA. Optimizing the Display and Interpretation of Data. Elsevier, Amsterdam, 2015:61-69.
4. Warner RA. Optimizing the analysis of clustered data. Proceedings of the 2012 International Conference on Bioinformation and Computational Biology. Edited by H.R. Arabnia and Q. Tran, CSREA Press, USA, 2012, p. 58-63.
5. Andresen A, Dalla Gasperina M, Myers R, Wagner GS, Warner RA and Selvester RH. An improved automated ecg algorithm for detecting acute and prior myocardial infarction. J. Electrocardiol. 35:105-110, Supplement 2002.
6. Warner RA, Hill N. Sheehe P, Mookherjee S, Fruehan. Improved criteria for the diagnosis of inferior myocardial infarction. Circulation 66:422-428, 1982.
7. Warner RA, Hill NE. Optimized electrocardiographic criteria for prior inferior and an anterior myocardial infarction. J. Electrocardiol. 45:209-213, 2012.
8. Warner RA. Optimizing the Display and Interpretation of Data. Elsevier, Amsterdam, 2015, p65, 68-69, Warner RA.
9. Warner RA. Optimizing the Display and Interpretation of Data. Elsevier, Amsterdam, 2015. pp 117-134.
10. Warner RA. Using the principles of bayesian statistics to improve the performances of medical diagnostic tests. Proceedings of the 2014 International Conference on Computational Science and Computational Intelligence. Edited by B Akhgar and H.R. Arabnia, IEEE Computer Society CPS, USA, 2014, p. 64-68.

Detecting Frequency from Randomly Sampled Data Implementation of random sampling in BRATUMASS

Luxi Li [1], Yizhou Yao[2], Meng Yao[1*], Erik D. Goodman[3], E. John R. Deller[4]
[1]School of Info Science and Tech, East China Normal University, Shanghai, China
[2]College of Science and Engineering, Central Michigan University, Mt Pleasant, MI, U.S.
[3]BEACON Center, Michigan State University, East Lansing, MI, U.S.
[4]ECE Michigan State University, East Lansing, MI, U.S.
[*]Corresponding Author, e-mail: myao@ee.ecnu.edu.cn

Abstract— *In this article, a system which implements random sampling theory is presented - the system obtains each sample after a random time interval, and it is a part of the Brest Tumor Microwave Sensor System(BRATUMASS) [1]; It is designed to refine the data of BRATUMASS. The first part introduces the system in the following aspects - the components of the system, how the signal is obtained, and the algorithm we used to calculate the spectrum of non-uniformly sampled data. The second part introduces a set of experimental performances based on random sampling method to explore the features of random sampling; the signals used in the experiment are single frequency sinusoidal waves, mixed sinusoidal waves, and a piece of bass music waves [2], since the random sampling method is a prototype and not integrated with BRATUMASS yet. The data from BRATUMASS is a uniformly sampled data interpolated with the same mechanism - random time interval interpolation, and it will be the next step of this study.*

Keywords: Compressive Sensing, Microwave Imaging, Random Sampling

1 Introduction

Breast cancer usually comes from breast tumor which could later become worse and convert to breast cancer. Thus the earlier breast cancer is detected, the more likely to practice permanent cure. BRATUMASS is developed to detect breast cancer at an earlier stage, more treatable stage. Thus, In BRATUMASS, the analysis of the data becomes significantly important. In BRATUMASS, the frequency resolution is such important in discriminating different breast tissue. Here is random sampling comes into the picture-random sampling is a method of compressive sensing [3]. It features a signal to information conversion. Due to limited of information energy in the Nyquist-Shannon theorem, random sampling reduces the number of samples without much perceptual loss. This quality is ideal in the developing of a portable cost-effective device. More importantly, the signal to information conversion, if successful, would increase the information to noise ratio while processing BRATUMASS's data and improve the imaging resolution of BRATUMASS. This work is to explore the features of random sampling and to prepare for its implementation in BRATUMASS.

2 Glance of the system

The Randomly Sampled Data system consists of sampling module, FIFO to USB converter, data display and analysis platform on PC.

A. Sampling Module: This module is based on an Arduino [4]. Timer, ADC module and random number generator mentioned below are all integrated in this chip. Timer is set according to a sequence of random number. ADC module is triggered by the compare match of the timer and the value of compare match register is updated in each of the compare match interrupt routine. Thus the time interval between two neighbouring sample is controlled by a random number generated by algorithm. Both ADC data and the random time interval is send to PC through FIFO to USB module.

B. FIFO to USB Converter: Stores data from the sampling module before computer reads it and sends message from computer to sampling module [5].

C. Data display and analysis platform on PC: Sends instructions to and reads data from the USB port. Display waveform and stores data for analysis.

3 Spectrum calculation

The anti-aliasing properties of random sampling make it possible to represent signal from a relatively small set of data. In this experiment, each sample is obtained after a random time interval. Simulation of this method has been done in previous work [6]. The advantage of this method is that aliasing can be avoided while sampling at an average rate below the Nyqvist rate. Besides, lower sampling rate means more observation time at a limited data amount. In other words, it takes longer to full a limited storage at a lower sampling rate. In this case, lower sampling rate makes longer observation time and thus increases resolution of the spectrum.

The difference of spectrum calculation between random sampled data and uniformly sampled data is the difference of time should take into account while doing integration [7].

Assuming that $x(t)$ is a band limited signal, $X_c(f)$ is Fourier transform of $x(t)$, sampling interval is T, total number of sample is N, then NT is the total sampling time. Let $x(n)$ be the uniformly sampled data, $x(t_n)\{n = 1,2,3,,n\}$ be the randomly sampled data, $X_D(f)$ be the Fourier transform of $x(t_n)$. We have:

$$X_c(f) = \int_0^{NT} x(t)e^{-j2\pi f t}dt \quad (1)$$

$$X_c(f) = \sum_{n=1}^{N} x(n)e^{-j2\pi f n} \quad (2)$$

$$X_D(f) = \sum_{n=1}^{N} x(n)e^{(-j2\pi f t_n)/(t_{n+1}-t_n)} \quad (3)$$

4 Random number generation

Random number of three kinds of distribution is used in this experiment; they are Uniform, Normal and Rayleigh distribution. These random numbers are interpreted as time interval between neighbouring samples. For comparison, Nyqvist style samples (identical time interval between neighbouring samples) is also taken and analysed. Therefore we have 4 groups of data for one target signal. HIST of a typical series of random number is shown in Fig. 1. These numbers are generated on data acquisition board and then transmit to PC.

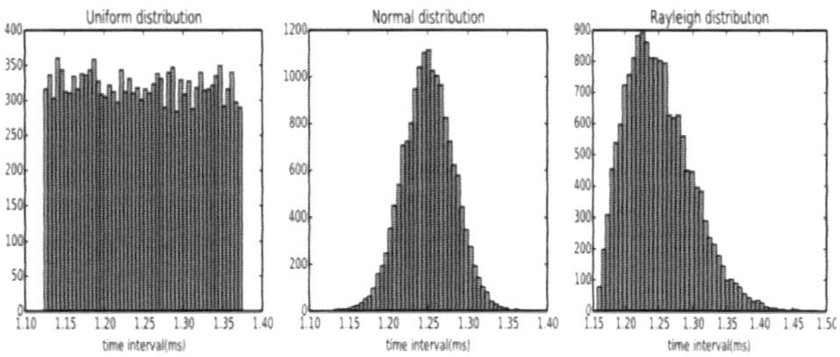

Fig. 1. HIST of random time interval

5 Signal and spectrum

Although perfect in simulation [6], practices in experiment shows limits while distinguish signal frequency at low sampling frequency. During the experiment, while sampling frequency is lower than the signal frequency, the spectrum of the signal failed to distinguish signal frequency. What caused this limit and how to improve this remains to be explored. However, distinguish of signal frequency lower than sampling frequency is successful. Experiments and results will be discussed in this part.

Five groups of signal are selected as target signal in this experiment. They are 233Hz, 678Hz, (233.5+233.6) Hz, (678.5+678.6) Hz sine waves and a piece of bass music wave. As is shown in Fig. 2, first we tried to distinguish a single sine wave, then a mix of two sine waves with close frequency, finally a bass music wave.

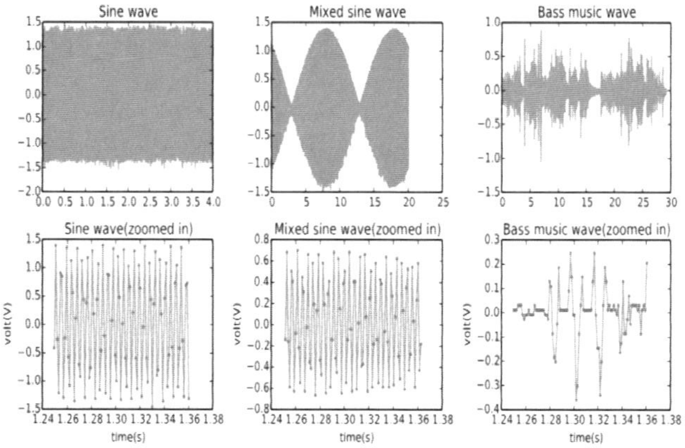

Fig. 2. Signal in time domain

5.1 Single sine wave

As is shown in Fig. 3, this sampling method is successful in distinguishing single sine wave. Aliasing is suppressed but frequencies with small amplitude are appearing where aliasing frequencies should be. This signal is obtained at an average sampling rate of 800Hz.

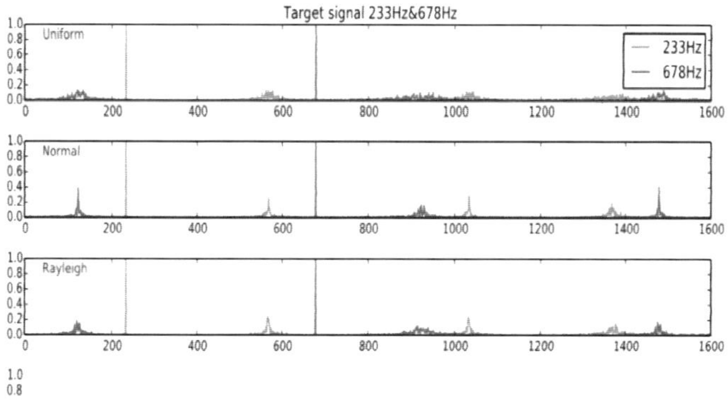

Fig. 3. Spectrum of single sine wave 233Hz and 678Hz in one figure

5.2 Mixed sine wave

In mixed sine wave experiment, a same amount of samples (16000 samples) is obtained from a mixed sine wave of 233.5Hz and 233.6Hz. The average rate of random sampling is 800Hz and the sampling rate of Nyquist sampling is 2000Hz. Thus the time window for random sampling is 20 seconds, and its frequency resolution should be 1/20=0.05Hz. While the time window for Nyquist sampling is 8 seconds, so the frequency resolution should be 1/8=0.125Hz. In other words, lower sampling rate means longer observation time window for signal when the data amount is limited. And the longer the observation time the higher the frequency resolution. The spectrum of this experiment shows that random sampling shows better frequency resolution at the same number of samples, as it is in Fig. 4 that random sampled data successfully distinguished mixed sine wave while Nyquist fails to distinguish two frequency components.

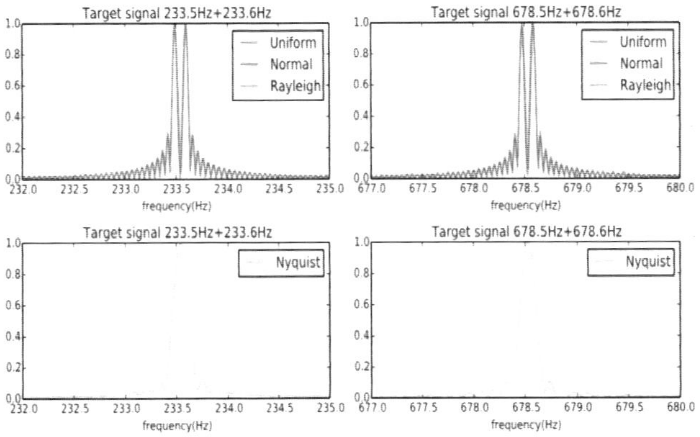

Fig. 4. Spectrum of mixed sine wave

5.3 Bass music wave

We also had a bass music wave experiment to explore how this mechanism fits sound wave with rich frequency components in it. This bass music wave is a wave(.wav) file with sampling rate of 44100Hz. Original bass music wave and its FFT of is showed in the last line in Fig. 5. Most of its energy is in frequency components within 1000Hz. All the sampling rate or the average of sampling rate in this bass music experiment is 800Hz.

The spectrum from random sampled data is no big difference from the signal's fft, except for the aliasing around 700Hz. The anti-aliasing propriety seems to be weakened when signal has rich main frequency components, shown in the Fig.5. The spectrum of Uniform, Normal, Rayleigh and Nyquist is roughly symmetric at an axis of 400Hz. All spectrum calculated from random sampled data is aliasing just like the spectrum calculated from Nyquist style sampled data. However the anti- aliasing propriety of random sampling is not totally gone. The amplitude of higher frequency, which is the aliasing frequency, is slightly lower than that of its symmetric frequency in Fig. 5.

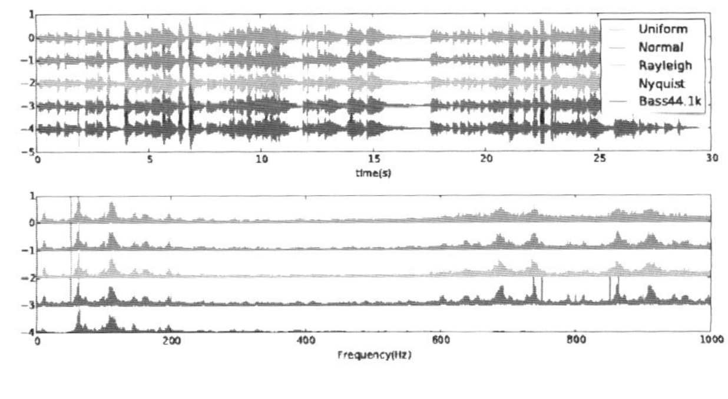

Fig. 5. Time and spectrum of bass wave

6 Refinement of BRATUMASS's data

A uniformly sampled data from BRATUMASS is interpolated with random time interval. The signal is obtained at 500Hz. It is interpolated at an average rate of 5000Hz. As is shown in Fig. 6, The samples interpolated between the original samples follows the linear relationship. Because the target signal in BRATUMASS is in 0-50Hz, the spectrum between 0 and 50Hz is compared in the third line of Fig. 6. Different deviation from 5000Hz in this experiment is to explore the influence of different σ while calculating the time interval of interpolation. The spectrum of signal after interpolation deviates from the spectrum before interpolation as the deviation grows. As is shown in Fig. 6 that the subtraction of the two fluctuates as the deviation grows. The result might be different while using larger σ or different mean. How different distribution of random numbers affect the signal spectrum is still to be explored.

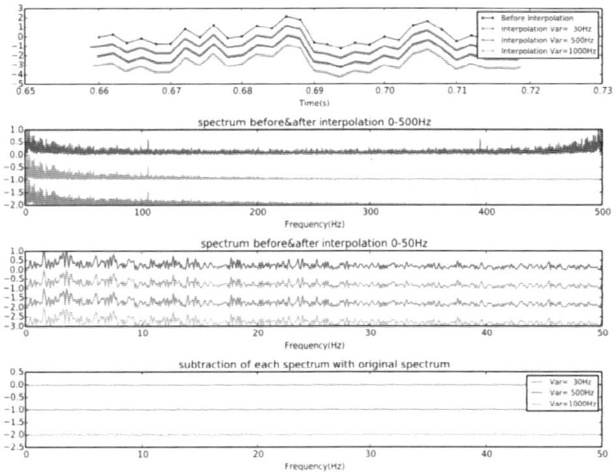

Fig. 6. Signal and its spectrum before and after interpolation

7 Conclusion

This Randomly Sampled Data system successfully distinguishes the spectrum of sine wave and bass wave. It shows advantage in improving the resolution of spectrum and in the suppressing of aliasing. However, how the distribution of random numbers affects the signal spectrum and how to relate the distribution of random interval and the information energy is still to be explored.

Acknowledge

This work has been performed while Prof. Meng Yao was a visiting Professor in Beacon Center, Michigan State University, thanks to a visiting research program from Prof. Erik D. Goodman. M. Yao would also like to acknowledge the support of Shanghai Science and Technology Development Foundation under the project grant numbers 03JC14026 and 08JC1409200, as well as the support of TI Co. Ltd through TI (China) Innovation Foundation.

References

[1] Z. Tao, "Investigation on the methodologies of near-field microwave echo imaging integrity," Ph.D. dissertation, East China Normal University, Shanghai, China, April 2011.

[2] freesound.org. (2012) double bass est1.[Online]Available: ttps://www.freesound.org/people/wescwave/sounds/169885/

[3] E. J. Candes and M. B. Wakin, "An introduction to compressive sampling," *IEEE SIGNAL PROCESSING MAGAZINE*, 2008.

[4] A. LLC.(2016)Arduino products.[Online]. Available: https://www.arduino.cc/en/Main/Products

[5] F. T. D. I. Ltd. (2015) Ft2232 - hi-speed dual usb uart/fifo ic. [Online]. Available: http://www.ftdichip.com/Products/ICs/FT2232H.htm

[6] Z. Cai, "Non uniform sampling used in breast cancer detection system," Master's thesis, East China Normal University, Shanghai, China, April 2015.

[7] A. Wang, "Methods of signal frequency measurement and implementation based on nonuniform sampling," Ph.D. dissertation, Huazhong University of Science and Technology, Wuhan, China, April 2004.

Research on approach for classification of Within imbalanced data sets

Chunkai Zhang[1], **Jiayao Jiang**[2], **and Fengxing Shi**[3]
[123]Department of Computer Science and Technology, Harbin Institute of Technology Shenzhen Graduate School, China
[1]ckzhang812@gmail.com
[2]616905919@qq.com

Abstract - *Most of the existing methods for unbalanced data classification only consider about the situation of imbalance between classes but don't consider about the situation within the class, thus affect the final classification results. In order to eliminate the imbalance within the class, put forward the cluster algorithms based on DBSACN algorithm to process the imbalance problem within the class. Through the determination of adaptive the ε and MinPts of DBSCAN algorithm, then form clusters and resampling within them. Then resolve the data fragmentation and density problems of the imbalance within class. Use UCI data for testing, and then compare the %ACC、F-Measure and AUC with other algorithms to prove the effectiveness of the algorithm.*

Keywords: clustering, data fragmentation, re-sampling

1 Introduction

Classification is an important task in pattern recognition, a series of classical classification algorithms, such as decision trees, neural networks, SVM, etc., have been developed and successfully applied to many important areas [1]. However, using the classic classifier to classify the unbalanced data will encounter many problems [2,3]. The characteristics of imbalanced data that is one kind of sample is much less than the others are, and in this case, we should pay more attention to this kind of sample. In reality there are many cases of imbalanced data set, such as satellite image classification, oil spill [4,5], medical diagnosis [6] and credit risk management [7,8].

In imbalanced data classification, there is a large difference in the number of samples in each class, which leads to the asymmetric information of the training algorithm. Because the traditional classifier is based on the maximization of the accuracy of the training, so often ignore information of minority class, but in many areas, the recall rate of the minority class is more important. Weiss mentioned that if the distribution of imbalanced data is not optimal [9], before the classifier is trained, it is needed to modify the sample space according to different evaluation criteria. Data resampling is a very effective method to solve the imbalanced data classification. It directly operates on the training set, changes the distribution of the training set, reduces the imbalance degree, and the processed samples are used to construct the classifier.

In this paper, we consider the class imbalance within class, and improve the classification results of the imbalance data set. This paper presents an improved DBSCAN algorithm. Through the determination of adaptive the ε and MinPts of DBSCAN algorithm, hen form clusters and resampling within them. Then resolve the data fragmentation and density problems of the imbalance within class.

2 Basic Concepts

2.1 SMOTE over-sampling algorithm

SMOTE algorithm is a popular over sampling technology [10]. It is mainly over sample surrounding minority class samples, rather than insert the copy data of the minority class samples. An algorithm that has been generalized from a project whose purpose is handwriting recognition mainly inspires this technique. The main process of SMOTE over sampling is as follows: first, according to each samples belongs minority class, select k nearest samples from the same minority class, and then do linear interpolation method between it and the k samples. SMOTE interpolation process is shown as the formula (1):

$$p_i = x + rand(0,1) * (y_i - x) \quad i=1,2,...N \qquad (1)$$

SMOTE algorithm is mainly doing interpolation between the close samples. Therefore, the over fitting problem can be avoided in the SMOTE algorithm, and the decision space of the minority class can be extended better. Similarly, it can also be applied to the majority class, which can reduce the decision space of the majority class. The pseudo code of SMOTE algorithm is as follows:

Algorithm 1. SMOTE algorithm pseudo code

Input: the number of samples of minority class, over sampling rate N%, nearest neighbor parameter K

Output: The new minority class

(1) *For i =1 to T*
Get K nearest neighbor of each minority samples, and save these samples to Karray

(2) *While N != 0*

(3) *SMOTE (N, i, Karray)* the function of SMOTE formula is, for the samples of minority class, according to the sampling rate, Linear interpolation is performed in Karray.

(4) *N = N - 1*

(5) *End while*

(6) *Return T**

In 2013, Lou Xiaojun proposed a new method based on clustering and took into account the boundary sample information. The method use unilateral selection method to determine boundary sample information of the minority class and the majority class. Cluster all minority class samples besides considering the boundary of the cluster, and over sampling in minority class. Accordingly, the replica of the minority class samples and the original sample has more similarity. And they can represent the space distribution of the parent class. Over sample from the boundary samples to increase sharpness of borders, thus emphasizing the importance of the boundary samples of minority class.

The idea of the under sampling method is to remove some specific samples according to some rule in majority class, and the sample data space can be balanced. However, there are some disadvantages of under sampling, which can easily cause the loss of important information of some representative samples.

Mixed sampling is an algorithm that combines over sampling and under sampling algorithm, in the process of resampling the original data set, using mixed sampling to balance the data set. A large number of studies have indicated that the mixed sampling method has more advantages than the single resampling method, and first oversampled then under sampling generally achieve better results. In this paper, the mixed sampling is adopted.

2.2 Borderline algorithm

In this section, we mainly describe the Borderline algorithm. The main function of the algorithm is to determine which belong to the boundary samples in the data samples, and the boundary samples are divided into the majority class samples and minority class samples. In this algorithm, we specify the imbalanced data set to be S, each of which is composed of the feature vector and the class label of the sample. The feature vector is x, the class label is y, that is, $x=\{x_1, x_2, \ldots, x_n\}$, y={Maj, Min}. So the data set can be expressed as:

$$S = \{(x_1, y_1)(x_2, y_2), \ldots, (x_n, y_n)\} \quad (2)$$

S in the majority of samples set is labeled as Smaj, also Smin represent a minority class. The process of the Borderline algorithm is as follows:

(1)To the Smin, find the K nearest neighbor samples in the whole data set S, and the samples are stored in the set KNNsmin corresponding to each Smin sample.

(2)To classify each of the samples in Smin into boundary samples, noise samples and safety samples by using the following three formulas:

$$\left|\{(x,y) | (x,y) \in KNN_{smin} \wedge y = Maj\}\right| > K/2 \quad (3)$$

$$\left|\{(x,y) | (x,y) \in KNN_{smin} \wedge y = Maj\}\right| = K \quad (4)$$

$$\left|\{(x,y) | (x,y) \in KNN_{smin} \wedge y = Maj\}\right| = 0 \quad (5)$$

Samples satisfy the equation (3) is considered the noise samples; Samples satisfy the equation (4) is considered the boundary samples; Samples satisfy the equation (5) is considered the safe sample.

(3) Do second-stage operation to the majority class.

By the above operation, we can get boundary samples of minority class and boundary samples of majority class of collection space, so that in the next mixed sampling process, we can focus on the processing of boundary samples, thus increasing the definition of boundary samples, improving the classification effect.

3 Improved DBSCAN algorithm

In the classification of imbalanced data samples, if there is imbalance in the data set within class, the traditional clustering algorithm cannot be used to cluster the data sample into desired clusters. Imbalance within class contains uneven distribution of minority class, data fragmentation and a disjunct distribution, so the transformation of the traditional methods is needed, improve the ability to deal with the imbalance samples.

DBSCAN algorithm cluster the sample data according to the density of the sample clustering. Compared with the traditional algorithm, it has the following advantages:

(1) Compared with the K-means algorithm, DBSCAN does not need to specify the number of clusters and the initial centroid;

(2) The cluster shape of DBSCAN is not changed obviously;

(3) According to the actual situation to determine the parameters of algorithm to reach the filter noise.

The traditional DBSCAN clustering algorithm can solve many problems that cannot be solved by other clustering algorithms, but the traditional DBSCAN algorithm has some disadvantages when faced to the imbalanced data set:

(1) Because of the uniform ε and Minpts of the DBSCAN algorithm, the clustering results of DBSCAN algorithm are often not the most ideal clustering results when face to the unbalanced datasets. Therefore, in the case of uneven distribution within the class, it is needed to use different ε and Minpts parameters to cluster, so that to get better results.

(2)When there is a class of unbalanced data fragmentation or small disjunct distribution, standard DBSCAN algorithm cannot consider this, the classifier is likely to consider them as noise.

In view of the above problems, we can use the improved DBSCAN algorithm to deal with these problems, and then solve the problem of imbalance within the class, the basic idea is as follows:

Firstly, a set of EPS values based on distribution density can be obtained by taking into account the uneven distribution density of the samples in the class. In the unbalanced data sets, the distance between the samples of each minority class and the other minority samples is different, that is, the distribution density is different. The calculation of the distribution density is measured by calculating the distance between one sample and the nearest K minority class samples to it. The method is as follows: calculate the average distance between one sample and the k nearest sample to it. Then get the relative density of each point.

Then cluster these distance samples, and get N clusters and N average distance. Arrange these N values in ascending order.

Lastly, these values were used as the DBSCAN algorithm threshold value from small to large. Some noise data is deleted when each time of clustering

By using the improved DBSCAN algorithm, we cannot only produce the minority class cluster, but also can solve the within class imbalance problem, data fragmentation and small disjunct problem.

4 Experimental

4.1 Evaluation criterion

Taking into account the special characteristic of the classification problem, when using the traditional evaluation criteria, it will cause the following problems. In order to obtain a higher overall accuracy, traditional classification methods treat all minority class samples as majority class samples, but for the minority class samples the classification accuracy is 0. In this case, the traditional evaluation system will no longer be appropriate to the unbalanced classification problem. Therefore, we need complex and comprehensive evaluation criteria. These standards mainly have two kinds, one kind is "the atomic standard", another kind is "the compound standard", and a large number of experiments have evidenced their reliability. In addition, the Receiver Operating Characteristic (ROC) has been widely used in the evaluation of unbalanced sample classification.

As shown in Table 1, it is the confusion matrix for the two-classification problem and classification of imbalanced samples. According to the statistical characteristics in confusing matrix and the relationship between them, we can accurately assess the classification results.

Table 1. Confusion matrix.

	Classified as positive	Classified as negative
positive	Correct positive *TP*	Wrong negative *FN*
negative	Wrong positive *FP*	Correct negative *TN*

Formula (6) to the formula (9) lists some of the commonly used atomic evaluation criteria for the classification of imbalanced samples based on the confusion matrix.

$$Accuracy = 1 - ErrorRate = \frac{TP + TN}{Pc + Nc} \quad (6)$$

$$\Pr ecision = \frac{TP}{TP + FP} \quad (7)$$

$$\mathrm{Re}\,call = \frac{TP}{TP + FN} \quad (8)$$

$$F\text{ - }Measure = \frac{(1+\beta)^2 \cdot \Pr ecision \cdot \mathrm{Re}\,call}{\beta^2 \cdot \mathrm{Re}\,call + \Pr ecision} \quad (9)$$

F-Measure is most often applied to the evaluation of imbalanced sample classification, as shown in the formula 9. The F-measure calculated from recall, precision and composite balance factor, when recall and precision have achieved a higher value, F-Measure will achieve results that are more satisfactory.

ROC curve (Operating Characteristics Curve Receiver) is proposed by Swets in 1988, then be widely used in many fields. To the ROC, FPRate is the X-axis, and TPRate is the Y-axis to build the space. By setting a threshold value, obtained a pseudo-positive rate and the true positive rate value, connect these scattered points and get the ROC curve.

In addition, the AUC (Area under the ROC curve) is proposed to evaluate the results, the larger the AUC, the better the result.

4.2 Experimental data set description

In this paper, we use the experimental data set of eight data sets, which have imbalance samples within class, and two imbalanced data sets, which are obtained from the UCI database. Table 4-2 describes the feature of the data sets; #Attr is the number of attributes that are included in the dataset; %Min is the proportion of minority class samples. The experiment consists of 10 data sets, each unbalance of which is not the same, and the total sample is not the same.

Table 2. Experimental data set description.

No.	Data-sets	#Attr.	%Min.
1	Yeast5	8	2.96
2	Abalone9	8	5.75
3	Glass1	9	8.85
4	Page-blocks0	10	10.23
5	Yeast4	8	10.98
6	Ecoli1	7	22.92
7	Vehicle3	18	25.06
8	Haberman	3	26.47
9	Glass0	9	32.71
10	Pima	8	34.90

4.3 Verification of improved DBSCAN clustering algorithm

This section is mainly to verify the effectiveness of the improved DBSCAN clustering algorithm by experiments.

The following experiments are mainly to verify the improved DBSCAN clustering effect and the difference between the clustering effect of ordinary clustering algorithm and it. K-means is chosen to represent the common clustering algorithms. Data set is a synthetic data, and the experimental results are shown below:

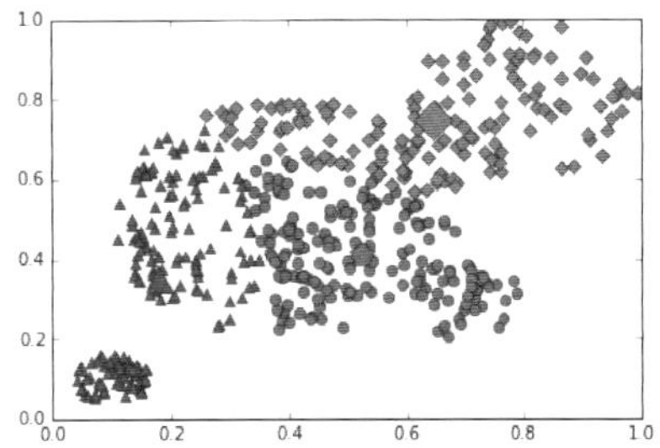

Figure 1. K-means clustering effect diagram

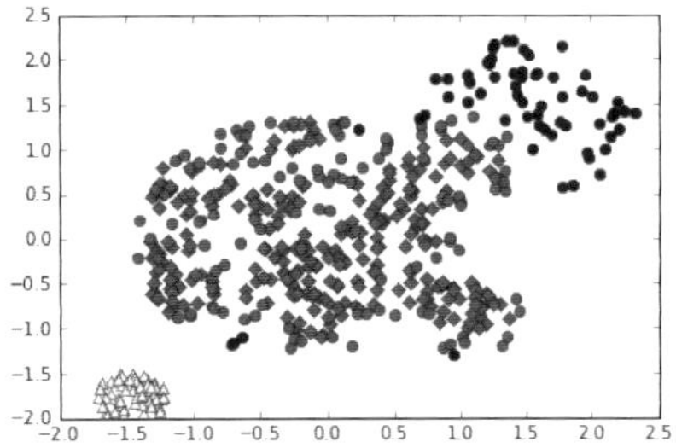

Figure 2. Improved DBSCAN algorithm clustering effect diagram

According to the above two clustering results, we can draw the following conclusion: the general clustering algorithm is affected by the initially cluster centroid and cluster number. we will not get the most ideal result, and the improved DBSCAN algorithm considering the imbalanced data set is not affected by artificial interference. The improved DBSCAN algorithm processing imbalance problem within class is better than the general clustering algorithm.

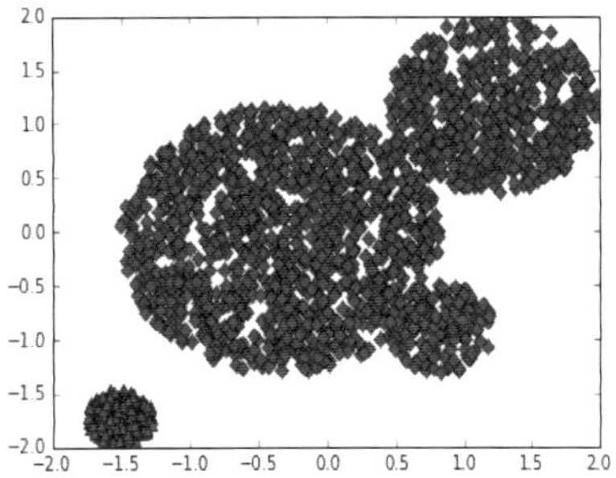

Figure 3. Improved DBSCAN-SMOTE over sampling effect chart

After clustering by the improved DBSCAN algorithm, oversampling is implemented, the result of SMOTE algorithm is shown in figure 3. As the figure shows that over sampling from on the cluster clustered by the improved DBSCAN algorithm will get the sample nearby the parent samples, so that these samples can represent parent samples.

The second experiment is use the ten UCI data sets to verify the effectiveness of the improved DBSCAN put forward in this paper. And the Comparison of K-means, DBSCAN and improved DBSCAN is in the follow table3.

The third experiment in this section is presented to verify the effectiveness of the improved algorithm DBS (DBSCAN-Smote) algorithm on the data level. In this experiment, the traditional data-resampling algorithm ROS (Random-Over Sampling), SM (SMOTE) and KBS (K-means-Borderline-SMOTE) are selected to compare with DBS.

Table 4 shows the result of the three algorithms on the ten UCI data sets. By compare F-Measure and Acc to evaluate the result is good or bad. Moreover, the classification algorithm is Adaboost.M1.

Table 3. Experimental results of the classification of three clustering algorithms

No.	%Acc		
	Kmeans	DBSCAN	Improved DBSCAN
1	95.68	96.38	**96.78**
2	89.84	91.54	**91.84**
3	82.29	89.63	**90.39**
4	**95.98**	95.48	95.28
5	91.77	92.28	**92.87**
6	87.61	87.58	**88.61**
7	74.28	77.67	**77.84**
8	66.3	70.9	**72.35**
9	80.52	82.71	**82.82**
10	72.09	70.74	**73.09**

Table 4. the F-Measure and %Acc values of the DBS algorithm

No	F-Measure			%Acc		
	ROS	SM	DBS	ROS	SM	DBS
1	0.731	0.730	**0.740**	**98.18**	97.11	97.93
2	0.313	0.344	**0.424**	90.70	85.91	**93.84**
3	0.250	0.280	**0.431**	86.98	75.00	**92.47**
4	0.853	0.838	**0.879**	**96.88**	96.35	96.69
5	0.759	**0.777**	0.743	94.07	94.47	**94.70**
6	0.810	0.771	**0.843**	91.07	91.07	**92.05**
7	0.524	**0.603**	0.558	76.00	76.36	**85.47**
8	0.475	0.468	**0.543**	66.01	65.03	**79.73**
9	0.653	**0.701**	0.685	76.64	77.10	**83.41**
10	0.601	0.596	**0.659**	70.97	69.92	**80.72**

It can be seen from table 4. DBS algorithm put forward in this achieved the highest F-Measure value 6 times out of 10, at the same time there is 7 times the highest overall accuracy rate, its performance is best. From these evaluation indexes, we can conclude that DBS is more effective than the existing traditional algorithm, which can effectively solve the effect of imbalance within class to the classification. The data set "yeast5" and "page-blocks0" only contain imbalanced data between classes, but not contain within class, and the DBS have a good performance.

Figure 4 and figure 5 show the experimental results F-Mea and Acc in histogram effect.

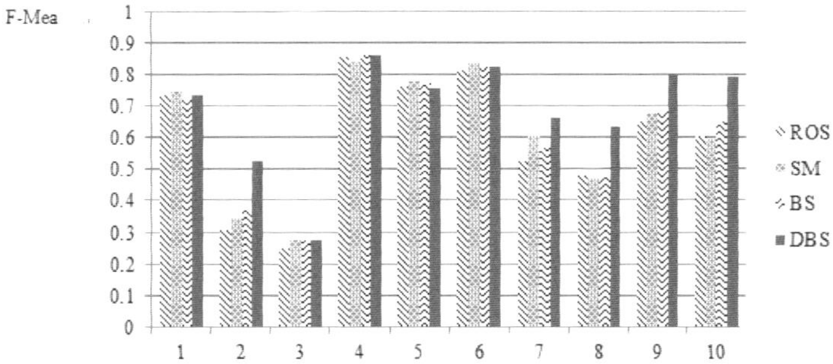

Figure 4. DBS algorithm F-Mea histogram

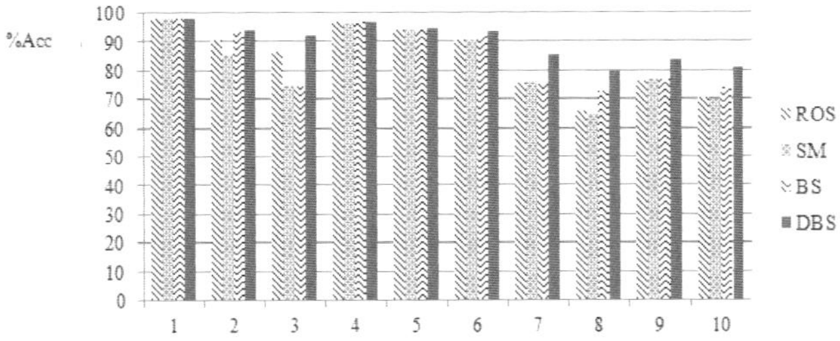

Figure 5. DBS algorithm %Acc histogram

These two figures show an important conclusion: when the data set is not balanced within the class, the more serious is the imbalance, the larger DBS algorithm to enhance the results of the classification. When DBS is applied to the data set with imbalance within class, final classification will enhance the effect of 6 % to 10%. It also fully proves that the DBS algorithm has obvious superiority in solving the problem of the classification of imbalanced samples within class.

The following figure6 shows the relationship between the improvement of the DBS and the degree of imbalance within class. The picture shows that the more imbalanced the data is the more improvement DBS has, compared with the traditional algorithm. It is further verified that DBS can reduce the impact of imbalanced data.

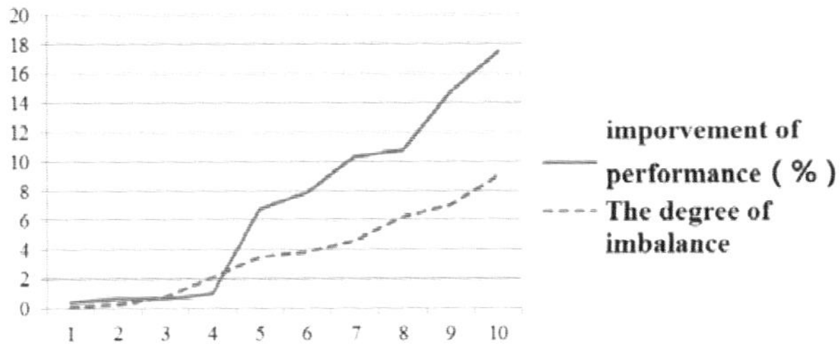

Fig. 6. Improve of algorithm and the degree of imbalance within class

5 Conclusions

When classify the imbalanced data, most of the traditional algorithm only consider the imbalance between classes, but ignore the imbalance within the class, so they will not get a satisfactory result. This paper mainly optimize the classification on the data level, and proposed solutions of classify the data with imbalance within the class. The DBS (DBSCAN – SMOTE) algorithm can over sample the data by considering the boundary information and local density of the samples. Finally we use 10 UCI data sets to prove the effectiveness of the algorithm. By contrasting the ACC and f-measure of the classification result with several algorithms, get the conclusion that the DBS can get the most satisfactory result in most of the data sets and a better result on the data set, which has the imbalance within class.

6 References

[1] Vandenberghe R, Nelissen N, Salmon E, et al. Binary Classification of F-flutemetamol PET Using Machine Learning: Comparison with Visual Reads and Structural MRI[J]. NeuroImage, 2013, 64:517-525.

[2] Zhai Yun, Ma Nan, Ruan Da, et al. An Effective Over-sampling for Imbalanced Data Sets Classification[J]. Chinese Journal of Electrics, 2011, 20(3):489-494.

[3] Lazarevic A, Ertoz L, Ozgur A, et al. Evaluation of Outlier Detection Schemes for Detecting Network Intrusions[C]//Proceedings of Third SIAM International Conference on Data Mining, 2003:97-104.

[4] Liu Y, Chen Y. Face Recognition Using Total Margin-based Adaptive Fuzzy Support Vector Machines[J]. Neural Networks, 2007, 18:178-192.

[5] Kubat M, Holte R C, Matwin S. Machine Learning for The Detection of Oil Spills in Satellite Radar Images[J]. Machine Learning, 1998:195-215.

[6] Yin L, Leong T. A Model Driven Approach to Imbalanced Data Sampling in Medical Decision Making[J]. Study Health Technology Information, 2010:856-860.

[7] Huang Y, Hung C, Jiau H. Evaluation of Neural Networks and Data Mining Methods on A Credit Assessment Task for Class Imbalance Problem[J]. Nonlinear Analysis: Real World Applications, 2006:720-747.

[8] Vaishali G. An Overview of Classification Algorithms for Imbalanced Data[J]. Emerging Technology and Advanced Engineering, 2012, 2(4):42-47.

[9] Jo T, Japkowicz N. Class Imbalances Versus Small Disjuncts[J]. ACM SIGKDD Explorations Newsletter, 2014, 6(1):40-49.

[10] Chawla N V, Bowyer K W, Hall L O, et al. SMOTE: Synthetic Minority Over-sampling Technique[J]. Journal of Artificial Intelligence Research, 2002: 321-357.

A Proposed Warped Choi Williams Time Frequency Distribution Applied to Doppler Blood Flow Measurement

F. García-Nocetti, J. Solano, F. and E. Rubio
Instituto de Investigaciones en Matemáticas Aplicadas y en Sistemas
Universidad Nacional Autónoma de México
Circuito Escolar, Cd. Universitaria, México City, 04510, México
fabian.garcia@iimas.unam.mx

Abstract - *One of the main goals in ultrasonic Doppler blood flow measurement is the estimation of the mean velocity. The Doppler signal's instantaneous frequency has traditionally been used to estimate the mean velocity. In this work, a non-uniform discrete time frequency distribution is proposed: the warped discrete Choi Williams distribution ($WTFD_{CW}$). The proposed procedure estimates the instantaneous frequency by concentrating the frequency resolution around the Doppler signal's instantaneous frequency. As a result, a better precision is obtained in the spectral estimation by using a $WTFD_{CW}$ for noisy signals when compared to other methods such as the Discrete Choi Williams Time Frequency Distribution ($DTFD_{CW}$) with the instantaneous frequency calculated as the centroid of the spectrum.*

Keywords: Signal Processing, Time-Frequency Distributions, Warped Fourier Transform, Doppler Flow Measurement.

1. Introduction

It is known that the blood flow mean velocity through a vessel's cross section is proportional to the instantaneous frequency of the Doppler ultrasonic signal. This work is focused on the accurate computation of the instantaneous frequency of a signal (Carotid Artery simulated signal) in the presence of noise.

A classic method to estimate the instantaneous frequency of a signal includes the computation of its spectrogram using a Short Time Fourier Transform (STFT). However, it assumes that the analysed signal is stationary and it compromises its temporal and frequency resolution. An alternative method is to use the Cohen Class Time Frequency Distributions that overcomes the stationary assumption. However, in some cases, it is desirable to increase only its frequency resolution around certain frequency of interest, for example, around the instantaneous frequency. That can be achieved if the length of the analysed discrete signal is increased but it also increases notably the computational cost. On the other hand, non-uniform discrete Fourier transforms are available such as the Warped Discrete Fourier Transform, which is able to achieve that task without having to increase the length of the analysed discrete signal.

The aim of the work presented in this paper is to incorporate a warped frequency scale (non-uniform) to the Cohen class of time-frequency distributions. We focus on the development of the warped discrete Choi Williams time-frequency distribution, although the procedure can be easily extended to other distributions. In previous works, a warped discrete Wigner-Ville time-frequency distribution [10] and a warped discrete Modified-B [11] have been proposed.

2. Time Frequency Distributions of the Cohen Class

The time frequency distributions of the Cohen class (TFD) are defined as follows [1]. Let $\phi(\theta,\tau)$ be the distribution kernel. That kernel determines the distribution and its characteristics of temporal and frequency resolution. Let also $\psi(t,\tau)$ be the Fourier transform of the distribution kernel:

$$\psi(t,\tau)=\frac{1}{2\pi}\int_{-\infty}^{\infty}\phi(\theta,\tau)e^{-it\theta}d\theta \quad (1)$$

Then, let

$$R_t(\tau)=\int_{-\infty}^{\infty}\psi(t-\mu,\tau)x(\mu+\tfrac{1}{2}\tau)x^*(\mu-\tfrac{1}{2}\tau)d\mu \quad (2)$$

be the deterministic generalised local auto-correlation function, where $x(t)$ is a complex signal. Finally, let

$$TFD(t,\omega)=\frac{1}{2\pi}\int_{-\infty}^{\infty}R_t(\tau)e^{-j\omega\tau}d\tau \tag{3}$$

be the time frequency distribution, which is defined as the Fourier transform of the local auto-correlation function, where t is the time variable and ω is the (angular) frequency variable.

3. Choi Williams Time Frequency Distribution

The Choi Williams time frequency distribution (TFD_{CW}) belongs to the Cohen class [9]. Its kernel is:

$$\phi(\theta,\tau)=e^{-\frac{\tau^2\theta^2}{\sigma}} \tag{4}$$

Then, the distribution is defined by:

$$TFD_{CW}(t,f)= \int_{-\infty}^{+\infty}\sqrt{\frac{1}{4\pi\tau^2/\sigma}}\int_{-\infty}^{+\infty}e^{-\frac{(t-\mu)^2}{4\tau^2/\sigma}}x\left(\mu+\frac{\tau}{2}\right)x^*\left(\mu-\frac{\tau}{2}\right)d\mu e^{-j2\pi f\tau}d\tau \tag{5}$$

where t is the time variable and f is the frequency variable. The direct discretization of expression (5) constitutes the discrete distribution ($DTFD_{CW}$), that is:

$$DTFD_{CW}(n,k)=2\bullet \sum_{p=-N+1}^{N-1}W^*(-p)W(p)\sum_{m=-M}^{M}\left(\sqrt{\frac{1}{4\pi p^2/\sigma}}e^{-\frac{m^2}{4p^2/\sigma}}\right)x^*(m+n-p)x(m+n+p)\left(e^{-\frac{j2\pi kp}{L}}\right)^2 \tag{6}$$

where n is the time discrete variable, k is the frequency discrete variable, $W(n)$ is a sampling window and $x(n)$ is a discrete complex signal with support $n=-N+1,\ldots,N-1$ and length $L=2N-1$.

Now, the discrete Choi Williams time frequency distribution with periodic extension ($PTFD_{CW}$) is stated [2]. The procedure essentially consists on the following. First the discrete distribution (6) is valued at $n=0$:

$$DTFD_{CW}(0,k)=2\bullet \sum_{p=-N+1}^{N-1}W^*(-p)W(p)\sum_{m=-M}^{M}\left(\sqrt{\frac{1}{4\pi p^2/\sigma}}e^{-\frac{m^2}{4p^2/\sigma}}\right)x^*(m-p)x(m+p)\left(e^{-\frac{j2\pi kp}{L}}\right)^2 \tag{7}$$

Second, the generalised local auto-correlation function is identified:

$$R_t(p)= W^*(-p)W(p)\sum_{m=-M}^{M}\left(\sqrt{\frac{1}{4\pi p^2/\sigma}}e^{-\frac{m^2}{4p^2/\sigma}}\right)x^*(m-p)x(m+p) \tag{8}$$

where $p=-N+1,\ldots,N-1$. Now, the function $\overline{R_t}(p)$ which constitutes the periodic extension of $R_t(\tau)$, is constructed as follows:

$$\overline{R_t}(p)=\begin{cases}R_t(p) & 0\le p\le N-1\\ 0 & p=N\\ R_t(p-2N) & N+1\le p\le L\end{cases} \tag{9}$$

where $p=0,\ldots,L$ and its length is $\overline{L}=L+1=2N$. Finally, note that (7) can be written as:

$$DTFD_{CW}(0,k)=2\sum_{p=0}^{\overline{L}-1}\overline{R_t}(p)\left(e^{\frac{-j2\pi kp}{\overline{L}}}\right)^2 \tag{10}$$

At this point, the following scaling in the frequency axis is carried out. It consists on reducing by half the frequency resolution. The result is denominated a discrete Choi Williams time frequency distribution with periodic extension ($PTFD_{CW}$):

$$PTFD_{CW}(0,k)=2\sum_{p=0}^{\overline{L}-1}\overline{R_t}(p)e^{\frac{-j2\pi kp}{\overline{L}}} \tag{11}$$

4. Warped Discrete Fourier Transform

The main characteristic of the warped discrete Fourier transform (WDFT) [3][4][5] is that it can concentrate the frequency resolution around a frequency of interest, since it possesses a non-uniform frequency resolution. Contrary to the conventional discrete Fourier transform (DFT) that possesses a uniform frequency resolution.

The warped discrete Fourier transform with a first order all-pass filter is defined as:

$$WDFT(k)=\sum_{p=0}^{\overline{L}-1}x(p)\left[\frac{\alpha^*+e^{\frac{-j2\pi k}{\overline{L}}}}{1+\alpha e^{\frac{-j2\pi k}{\overline{L}}}}\right]^p \tag{12}$$

where $\alpha=|\alpha|\exp(j\varphi)$ is a complex parameter that determines the warped frequency scale, $x(n)$ with $n=0,\ldots,\overline{L}-1$ is a complex discrete signal with length $\overline{L}$, and $k=0,\ldots,\overline{L}-1$ is an index related to the discrete frequency.

The warped frequency scale mapping is given by:

$$\Omega_W = \Omega + 2\arctan\left(\frac{|\alpha|\sin(\varphi-\Omega)}{1+|\alpha|\cos(\varphi-\Omega)}\right) \quad (13)$$

where $-\pi \le \Omega \le \pi$ with $\Omega = 2\pi k/\overline{L}$ is the conventional uniform frequency scale.

The magnitude of the parameter α, $|\alpha|$, is selected according to the percentage of frequency points to be concentrated inside the spectral lobe related with the frequency of interest. The angle of the parameter α, φ, is selected according to the value of the frequency of interest.

5. Warped Discrete Time Frequency Distributions

In this work, a warped frequency scale is incorporated to the discrete time frequency distributions with periodic extension. Although this procedure is illustrated for the Choi Williams distribution, its generalisation can be directly generated. The procedure essentially consists on the following: to calculate the warped discrete Fourier transform of the periodic extension of the generalised local auto-correlation function, instead of calculating its conventional discrete Fourier transform. Then, the warped discrete Choi Williams time frequency distribution ($WTFD_{CW}$) is:

$$WTFD_{CW}(0,k) = 2\sum_{p=0}^{\overline{L}-1}\overline{R}_t(p)\left[\frac{\alpha^* + e^{\frac{-j2\pi k}{\overline{L}}}}{1+\alpha e^{\frac{-j2\pi k}{\overline{L}}}}\right]^p \quad (14)$$

where $k = 0,\ldots,\overline{L}-1$ and the signal $\overline{R}_t(p)$ is the periodic extension of the generalised local auto-correlation function (9).

6. Frequency Estimation using the Warped TFD

The procedure to estimate the instantaneous frequency of a signal with a dominating single frequency and a narrow bandwith follows [5]. First, the spectrum of the signal is calculated, using a conventional discrete Fourier transform:

$$S(k) = \left|\sum_{p=0}^{L-1} x(n)e^{\frac{-j2\pi kn}{L}}\right|^2 \quad (15)$$

Second, the value of the parameter $\alpha = |\alpha|\exp(j\varphi)$ is calculated. For this, the preliminary instantaneous discrete frequency of the signal (the centroid of the spectrum) is calculated:

$$k_i = \frac{\sum_{k=-N+1}^{N-1} k \bullet S(k)}{\sum_{k=-N+1}^{N-1} S(k)} \quad (16)$$

The angle φ of the parameter α is the instantaneous discrete frequency of the signal but expressed in radians. It is important to consider the reduccion by half of the frequency resolution.

Third, the warped discrete time frequency distribution of the signal is calculated according to (14). Then, the definitive instantaneous frequency of the signal is the phase associated to the frequency component with the maximum magnitude that is obtained.

$$\Omega_k = angle\left[\max_k \left\{\left|WTFD_{CW}(0,k)\right|\right\}_{k=0}^{\overline{L}-1}\right] \quad (17)$$

7. Application to Doppler Flow Measurement

It is known that the mean velocity of the blood flow through the cross section of a vessel is proportional to the instantaneous frequency of a Doppler ultrasonic signal. In this work, a femoral artery signal is considered for this study. The simulation of that signal is detailed in [6][7][8]. The theoretical instantaneous frequency is shown in figure 1. The sampling frequency used is 12800 Hz. Sampling windows with 50% of overlapping and lengths equal to 128, 256, 512, 1024 and 2048 are used.

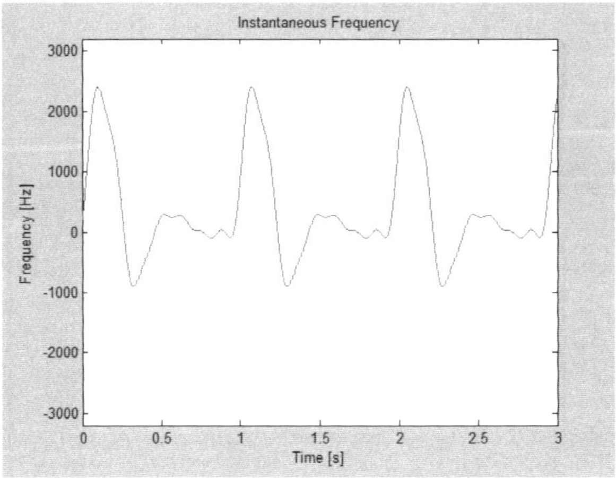

Figure 1. Theoretical instantaneous frequency of the simulated signal (Femoral artery).

For those conditions, the instantaneous frequency is calculated using both, the discrete Choi Williams time frequency distribution with periodic extension, $PTFD_{CW}$ (11), and the warped discrete Choi Williams time frequency distribution, $WTFD_{CW}$ (14). The procedure to calculate the instantaneous frequency has been described in the section 6. Finally, the RMS error respect to the theoretical instantaneous frequency is calculated.

8. Results

Figure 2 depicts a graph with the estimated instantaneous frequency using both the periodic extension and the warped discrete Choi Williams time frequency distribution.

Figures 3 show the RMS errors in the estimation of the instantaneous frequency for the simulated signal using different window lengths. Different levels of normalised gaussian noise (SNR of 40, 30, 20, 10 and 6 dB) have been added, with a concentration of frequency points around the instantaneous frequency of 40%. The precision of the warped distribution to estimate the instantaneous frequency is notoriously better in the presence of noise.

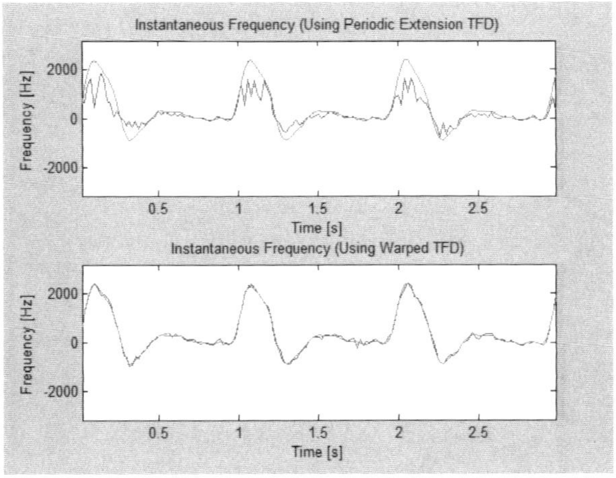

Figure 2. Estimated instantaneous frequency of the simulated signal using the periodic extension and the warped discrete Choi Williams time frequency distribution (L=512, SNR=10dB).

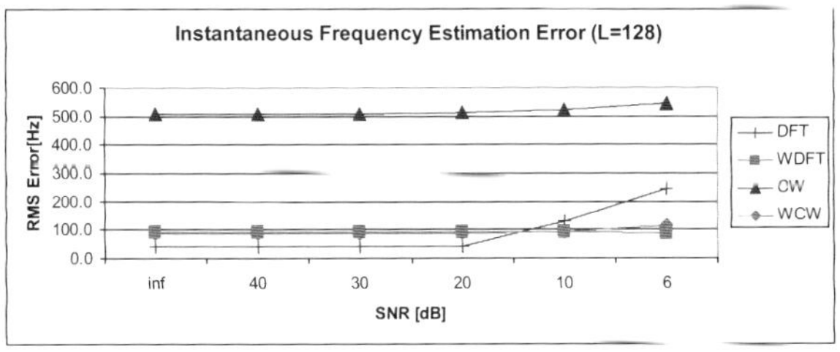

a)

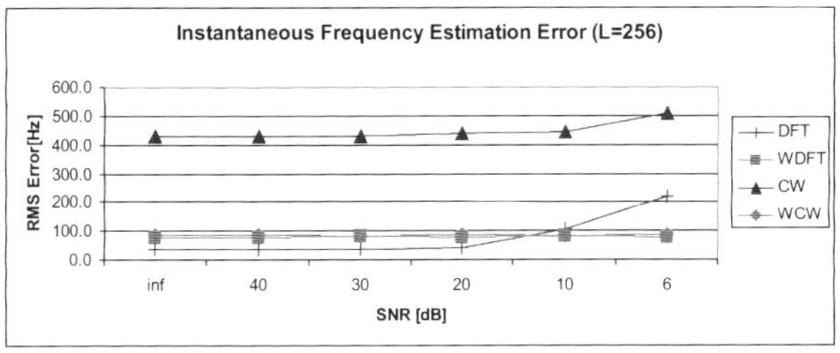

b)

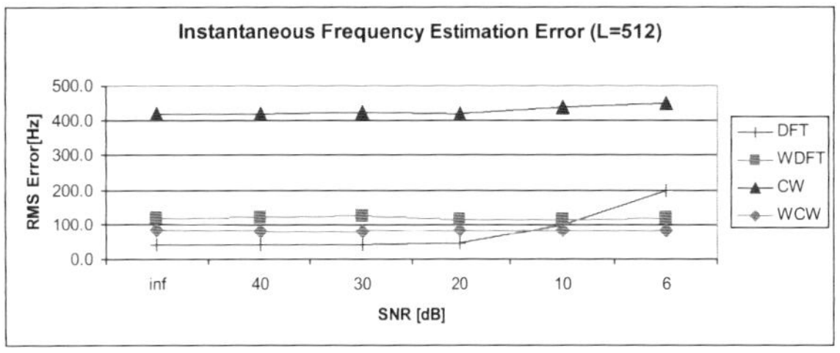

c)

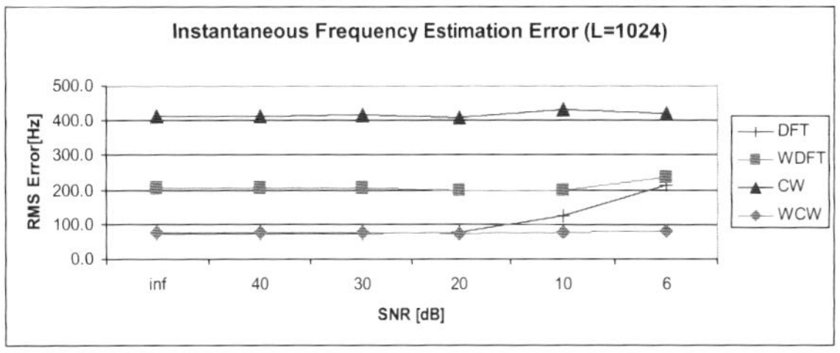

d)

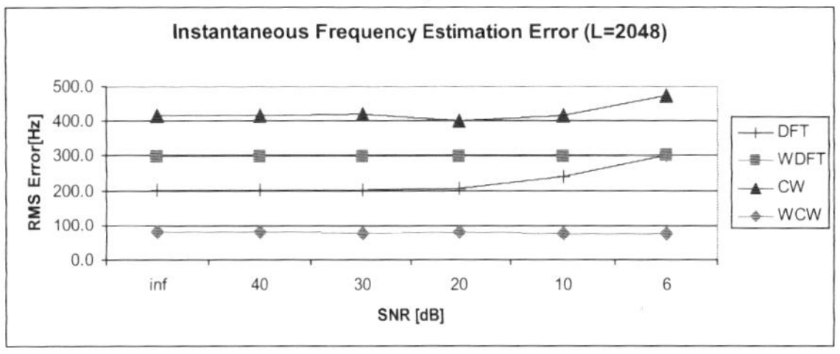

e)

Figure 3. RMS error [Hz] obtained in the estimation of the instantaneous frequency of the simulated signal. a) L=128, b) L=256, c) L=512, d) L=1024, e)L=2048.

9. Conclusions

The approach presented in this paper incorporates a warped frequency scale (non-uniform) to the Cohen class time frequency distributions. Particularly, the warped discrete Choi Williams time frequency distribution has been developed, although this procedure can be extended easily to other distributions. The method is applied to estimate the mean velocity of the blood flow through a vessel, which is proportional to the instantaneous frequency of the ultrasound signal obtained in the process of Doppler flow measurement.

The experiments have been carried out using a Femoral artery simulated signal. The results obtained by the warped discrete time frequency distribution are compared with those obtained by the discrete Choi Williams time frequency distribution with periodic extension. Results show that the distribution with periodic extension is easier to calculate since FFT-like algorithms of complexity $O(N \log N)$ are used; while the warped distribution uses algorithms which are based on the matrix multiplication whose complexity are $O(N^2)$.

Nevertheless, the precision of the warped distribution to estimate the instantaneous frequency is notoriously better in the presence of noise.

Acknowledgements

The authors acknowledge project DGAPA-UNAM PAPIIT (IN101213), project Consorciado CYTED (P506PIC0295) by the financial support. Also we want to acknowledge to M. Fuentes, J. Contreras, S. Padilla and M. Vazquez for their technical support in the development of this work.

References

[1] L. Cohen, *Time-Frequency Analysis* (Prentice-Hall PTR, 1995).

[2] B. Boashash, P. Black, An Efficient Real-Time Implementation of the Wigner-Ville Distribution, *IEEE Transactions on Acoustics, Speech, and Signal Processing, ASSP-35*(11), 1987, 1611-1618.

[3] A. Markur, S.K. Mitra, Warped Discrete Fourier Transform: Theory and Applications, *IEEE Transactions on Circuits and Systems -I:Fundamental Theory and Appications*, *48*(9), 2001, 1086 –1093.

[4] S. Franz, S.K. Mitra, J.C. Schmidt, G. Doblinge, Warped Discrete Fourier Transform: a New Concept in Digital Signal Processing, *Proceedings IEEE International Conference on Acoustics, Speech, and Signal Processing*, *2*, 2002, 205-208.

[5] S. Franz, S. K. Mitra, G. Doblinger, Frecuency Estimation using Warped Discrete Fourier Transform, *Signal Processing*, *83*, 2003, 1661-1671.

[6] J. Cardoso, G. Ruano, P. Fish, Nonstationary Broadening Reduction in Pulsed Doppler Spectrum Measurements Using Time-Frequency Estimators, *IEEE Transactions on Biomedical Engineering, 43*(12), 1996, 1176-1186.

[7] J. A. Jensen, *Estimation of Blood Velocities using Ultrasound* (Cambridge University Press, 1996).

[8] P. Fish, *Physics and Instrumentation of Diagnostic Medical Ultrasonic* (John Wiley Sons, 2000).

[9] H. Choi, W. Williams, Improved Time-Frequency Representation of Multicomponent Signals Using Exponential Kernels. *IEEE Transactions on Acoustics, Speech and Signal Processing, 37*(6), 1989, 862-871.

[10] E. Rubio, J. Solano, F. Torres, F. García-Nocetti, A Proposed Warped Wigner-Ville Time Frequency Distribution Applied to Doppler Blood Flow Measurement, *Proceedings of Biomedical Engineering (BioMED),* 2006.

[11] F. García-Nocetti, J. Solano, E. Rubio, A Proposed Warped Modified-B Time-Frequency Distribution Applied to Doppler Blood Flow Measurement, *Proceedings of the International Conference on Bioinformatics & Computational Biology (BIOCOMP),* 2015.

SESSION

SYSTEMS BIOLOGY AND RNA SEQUENCE DATA PROCESSING + HIGH PERFORMANCE COMPUTING

Chair(s)

TBA

Stage-specific transcriptome of the malaria parasite in the red blood cell cycle

Hao Zhang[1], Timothy G. Lilburn[2],Hong Cai,[1], Yufeng Wang [1, 3*]
[1] Department of Biology, University of Texas at San Antonio (UTSA), San Antonio, TX 78249, USA.
[2] Novozymes NA, Durham, NC 27709, USA.
[3] South Texas Center for Emerging Infectious Diseases, UTSA, TX 78249, USA
Email addresses: HZ: hao.zhang@utsa.edu, TGL: TLR@novozymes.com, HC: hong.cai@utsa.edu, YW: yufeng.wang@utsa.edu, * Corresponding author

Abstract— The World Health Organization has estimated that in 2015 there were 214 million cases of malaria that led to approximately 438,000 deaths. The development of novel therapeutics largely relies on a better understanding of parasite biology and pathogenesis. Our analyses on time-series RNA-Seq data identified stage-specific genes expressed across the blood-stage in the malaria parasite, *Plasmodium falciparum*: in the ring and the early trophozoite stage, genes associated with hemoglobin degradation and transcription were highly expressed; in the trophozoite and early schizont stage, genes involved in glycolysis, the TCA cycle, mitochondrion organization, deoxyribonucleotide metabolic processes as well as DNA replication were upregulated; in the schizont stage, genes associated with merozoite invasion and actin filament organization were over-expressed. Our results revealed an orchestrated transcriptional machinery and a "just-in-time" mechanism for transcriptional regulation in the blood stage of the malaria parasite.

Keywords-malaria; systems biology; RNA-Seq; Plasmodium falciparum

I. Introduction

Malaria is a severe global infectious disease that causes fever, severe anemia, cerebral malaria and, if untreated, death. The infection is transmitted by an infected female *Anopheles* mosquito. Despite that approximately 3.3 million global lives were saved and the malaria death rates in Africa were cut nearly in half from 2000 through 2012, as a result of large-scale malaria prophylaxis and treatment interventions by the World Health Organization (WHO), malaria remains a major public health problem. According to the latest WHO report in 2015, there were 214 million cases and estimated 438,000 deaths due to malaria [1]. The vast majority of the deaths occur in children under age 5 in sub-Saharan areas.

The causative agents of malaria are *Apicomplexan* pathogens in the genus *Plasmodium*. *P. falciparum*, predominant in Africa, is the most deadly species of the five human malarial parasites. The complex lifecycle of malarial parasite poses a particular challenge to the study of malaria biology [2]. *Plasmodium* sporozoites inoculated by infected mosquitoes enter into the human bloodstream and quickly invade liver cells, within which they differentiate into thousands of merozoites. The intraerythrocytic developmental cycle (IDC) of malarial parasites initiates with the invasion of erythrocytes by hepatic merozoites and is followed by a 48-hour cycle of asexual replication involving the ring stages, the trophozoite stage, and the schizont stage. Mature schizonts rupture to release newly formed merozoites that lead to reinvasion of new red blood cells (RBCs).

In spite of a 37% decline in the incidence of malaria in the last 15 years, propelled by a global fight against malaria that relies on insecticide-treated mosquito nets, indoor residual spraying, and new synthetic drugs, the rapid evolution of drug-resistant parasites and the lack of licensed vaccines still leaves millions of people suffering from this disease. This spurs the rush to develop new antimalarial drugs. A significant barrier to the search for novel drug targets is our lack of understanding of the parasite biology throughout its dynamic life cycle. Thanks to the development of high throughput RNA-Seq technology [3], a high-resolution transcriptomic landscape is available. Here we report our analysis of the time-series RNA-Seq data from the IDC of the malaria parasite. The red blood cell cycle was chosen as the focus of the study, as all the clinical symptoms are manifested in this cycle. A better understanding of transcriptional regulation will shed more light on the molecular mechanisms underlying parasite survival and pathogenesis.

II. Methods

A. Data preparation and quality control

Fastq files of RNA-seq data collected at seven time points (0h, 8h, 16h, 24h, 32h, 40h, and 48h) during the red blood cell cycle of *P. falciparum* [3] were downloaded from the PlasmoDB database [4] (Accession number ERP000069). Sequences were trimmed using the modified-Mott trimming algorithm as implemented in the CLC Genomics Workbench 8.0 (Qiagen) with a limit parameter of 0.02. Hierarchical clustering (HC) and principal component analysis (PCA) were conducted for quality control.

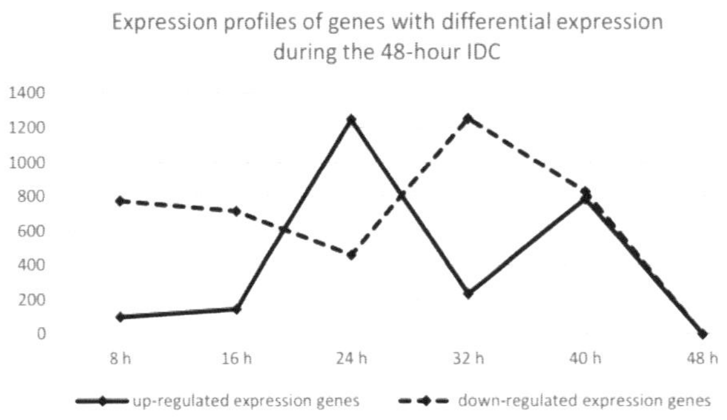

Figure 1. The number of genes with differential expression across the 48-hour IDC.

B. RNA-Seq analysis

All sequence reads were only mapped to the gene regions of the latest reference genome sequence of *Plasmodium falciparum*, PF3D7v3.0. The default values of mapping parameters were used except for the "Maximum number of hits for a read" which was set to 1. Expression value was defined as the unique counts of reads that mapped to the exon-regions of the gene. Differential expression analysis was conducted for pair-wise comparisons across the seven time points using the Exact Test as implemented in the EdgeR Bioconductor package [5, 6]. The Bonferroni method and the Benjamini and Hochberg's algorithm were used for multiple testing correction. The differential gene expression was considered statistically significant if the absolute value of the expression fold change ≥2 and the false discovery rate (FDR) corrected p-value ≤0.05.

C. Gene ontology (GO) enrichment analysis

In order to identify over-represented functional categories, GO enrichment analysis [7] was conducted in both the top 25 percentile of all the expressed genes and the significantly differential expression genes, with the cut-off p-value of 0.05. The Bonferroni method and the Benjamini and Hochberg's algorithm were used for multiple testing correction.

III. Results

A. Summary statistics of RNA-Seq data of the blood-stage P. falciparum transcriptome

Using CLC Genomics Workbench Version 8.0, we analyzed the RNA-Seq data collected from seven time points during the 48-hour IDC of *P. falciparum* [3]. Therefore, the expressional features of the entire RBC cycle, from the ring stage to the formation of new merozoites, were captured. We mapped the RNA-Seq reads to the latest reference genome sequence of *P. falciparum*-PF3D7v3.0, a completed genome published by Sanger Institute with no gaps and comprehensive reannotation. Among the 5,369 genes in the genome, we were able to detect the transcription of ~5,000 genes during the IDC. Two different strategies, 37 bp and 54 bp pair-end sequencing, were applied in transcriptome sequencing. More than 72% of the reads from the ring stage to the trophozoite stage were mapped to the reference. The number of mapped reads, however, from the schizont stage was lower than 70%, likely because a relatively higher percentage of low-complexity sequences was observed in the end of the IDC.

Using the Exact Test as implemented in the EdgeR Bioconductor package for pair-wise comparisons of mRNA expression across the seven time points, we found that, in the ring stage, 101 genes and 774 genes were up-regulated and down-regulated, respectively; 146 genes were highly expressed and 714 genes were expressed at lower level during the transition from the ring to the early trophozoite stage; however, 1,247 genes were activated and 458 genes were repressed at the trophozoite stage; during the transition from the trophozoite to the early schizont stage, the number of genes with increased expression dropped back to 236 but the number of genes with decreased expression increased to 1,254; in the schizont stage, the numbers of up-regulated and down-regulated genes were similar. Figure 1 shows the changes in the number of genes with differential expression during the 48-hour red blood stage, suggesting the expression profiles of the intraerythrocytic *Plasmodium* parasite were clearly dynamic: gene transcription was repressed in the ring stage and was maintained at the similar level during the transition phase from ring to early trophozoite stage, then, gene expression was activated significantly in the trophozoite stage; after that, the gene expression returned to an inhibitory state during the transition phase from trophozoite to early schizont stage; in the end, it turned to a balanced state of gene expression. We will discuss stage-specific expression in the following subsections.

B. The ring stage and the transition from the ring to the early trophozoite stage

After the invasion of RBCs, more than 4,000 genes were detected to be expressed during the ring stage (8h) and the transition from the ring to the early trophozoite stage (16h). The Gene Ontology (GO) enrichment analysis of the top 25% genes showed that the genes involved in hemoglobin catabolic process, retrograde vesicle-mediated transport from Golgi apparatus to ER, protein import into nucleus, nucleosome assembly, response to drug, glycolytic process, regulation of transcription, RNA splicing, regulation of translation and cell-cell adhesion were highly expressed in the ring stage.

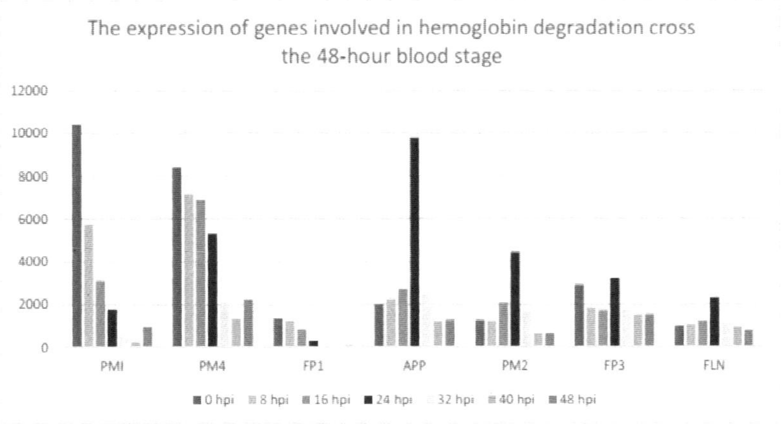

Figure 2. The expression of genes involved in hemoglobin degradation across the 48-hour blood stage.

The proteases regulating hemoglobin degradation, including plasmepsin I (PM1), plasmepsin IV (PM4) and cysteine protease falcipain 1 (FP1), had peak expression at the beginning of the IDC, then gradually decreased; on the other hand, other hemoglobin degradation-associated proteases, such as aminopeptidase P (APP), plasmepsin II (PM2), cysteine protease falcipain 3 (FP3) and falcilysin (FLN), were highly expressed in the ring stage and peaked at 24 hour post invasion (hpi), then repressed in the schizont stage (Figure 2). These proteases were shown to play an important role in hemoglobin digestion that is essential for the intraerythrocytic development of malaria parasite [8-11]. The end product of hemoglobin proteolysis is free heme, which is toxic to the parasite. Notably, the expression of heme detoxification protein (HDP), a potent protein that is capable of converting toxic heme into hemozoin [12], was not detected in the ring stage. It remains largely unknown how the parasite detoxifies heme; possible mechanisms may involve degradation facilitated by hydrogen peroxide in the food vacuole, or a glutathione-dependent degradation in the parasite's cytoplasm.

The genes regulating transcription were also significantly enriched in the ring stage. While it is believed that transcriptional regulation in the malaria parasite is a concerted complex process, little is known about the components as well as the control mechanism of transcriptional regulation [13]. We found that all three subunits of RNA polymerase (I, II, and III) were expressed at high level, and were among the top 25% highly expressed genes. In addition to basal transcription factors, 22 transcriptional regulators were induced during the ring stage. Fifteen of these 22 transcription factors were members of the ApiAP2 family with one of more characteristic AP2 (Apetala2) DNA-binding domain(s). Interestingly, AP2-L, which was shown to be required for the development of the rodent malaria parasite *P. berghei* in the liver-stage [14], and AP2-O, a major regulator of ookinete genes in multiple species in the genus of *Plasmodium* [13], were also highly expressed, suggesting that they may play additional roles in the ring stage of *P. falciparum*. Moreover, other DNA-binding protein, including a Myb2 transcription factor, and two high mobility group proteins (HMGB1 and HMGB2), were enriched. Several proteins that may be associated with transcriptional regulation were also expressed at a high-level, including transcriptional activators, bromodomain protein 1 (BDP1) which is required to coordinate the expression of invasion-related genes [15], as well as histone acetyltransferase GCN5 and a transcriptional coactivator ADA2. They were shown to interact with each other to promote transcriptional activation [16]. By contrast, histone deacetylase HDA1, a global transcriptional repressor, and AP2-G2, which functions as a repressor in gametocytogenesis [17], were also highly expressed. They may contribute to the prevalence of transcriptional repression present in the ring stage (774 genes down-regulated vs. 101 genes up-regulated). The regulation of gene expression also occurred at the level of protein translation. The top 20% of highly expressed genes in the ring stage included 120 genes associated with translation, including 18 transcription initiation factors and all of the 20 aminoacyl-tRNA synthetases.

C. The trophozoite stage and the transition from the trophozoite to the early schizont stage

In the trophozoite stage, the expression of genes involved in the cellular biogenic amine biosynthesis, including S-adenosylmethionine decarboxylase/ornithine decarboxylase (AdoMetDC/ODC), spermidine synthase, putative ribose-phosphate pyrophosphokinase and a conserved *Plasmodium* protein (PF3D7_0409300), peaked at 24 hpi, and then reduced immediately. The ethanolamine kinase (EK), however, was up-regulated significantly at the 24-hour time point and maintained a high-level of expression in the schizont stage. This observation is consistent with the report that in *P. falciparum* EK is a critical enzyme in the *de novo* biosynthesis of phosphatidylethanolamine, which is essential for parasite survival [18]. EK is therefore considered as a potential antimalarial drug target.

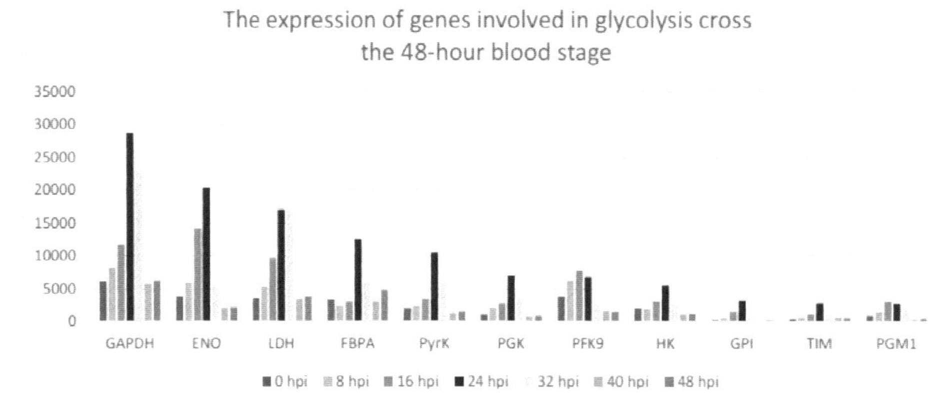

Figure 3. The expression of genes associated with glycolysis across the 48-hour blood stage.

The enzymes associated with carbohydrate metabolism were active in the trophozoite stage. The genes involved in glycolysis were highly expressed at the ring stage and most of them were decreased at 32 hpi after peaking at 24 hpi (Figure 3); this is consistent with previous studies, indicating that glucose fermentation is a major energy source for intraerythrocytic *Plasmodium* parasites [19, 20]. The function of the tricarboxylic acid cycle (TCA cycle) in the blood-stage malaria parasites has long been considered enigmatic. Genome sequencing revealed that the *Plasmodium* genome encodes homologs to all the enzymes required to a complete TCA cycle [21]. The role of TCA in intraerythrocytic *Plasmodium* development was proposed to be minor given the microaerophilic environment required for an optimal development of parasites *in vitro* [22] and the predominant role of anaerobic glycolysis to energy generation. Our RNA-Seq analysis, however, revealed that the enzymes catalyzing steps in the TCA cycle, including putative citrate synthase (CS), a pace-making enzyme localized in mitochondrion responsible for the first step of TCA cycle, putative 2-oxoisovalerate dehydrogenase subunit beta, mitochondrial (BCKDHB), putative succinyl-CoA ligase (SCS) and putative succinyl-CoA synthetase alpha subunit (SCS α subunit), were activated at 32 hpi, immediately after the up-regulation of glycolytic enzymes (Figure 4). Also, malate:quinone oxidoreductase (MQO), an alternative to the malate dehydrogenase enzyme that catalyzes the reversible conversion of malate into oxaloacetate, and isocitrate dehydrogenase (IDH), were highly expressed at both 24 hpi and 32 hpi (Figure 4). Unlike other eukaryotes which possess at least three isoforms of IDH (mitochondrial NADP-dependent, mitochondrial NAD-dependent and cytosolic NADP-dependent), *P. falciparum* has only one isoform (mitochondrial NADP-dependent enzyme) of IDH [20]. In addition, the genes involved in mitochondrion organization and ubiquinone biosynthetic process, an essential metabolic function served by mitochondrion, were enriched at 24 hpi, suggesting the function of mitochondria is synchronized with the induction of TCA metabolism in the trophozoite stage.

In addition to genes involved in glycolysis and TCA, genes associated with deoxyribonucleotide metabolism, including bifunctional dihydrofolate reductase-thymidylate synthase (DHFR-TS), a validated target for antifolate antimalarials, deoxyuridine 5'-triphosphate nucleotidohydrolase (dUTPase), an enzyme that catalyzes the conversion of dUTP to dUMP

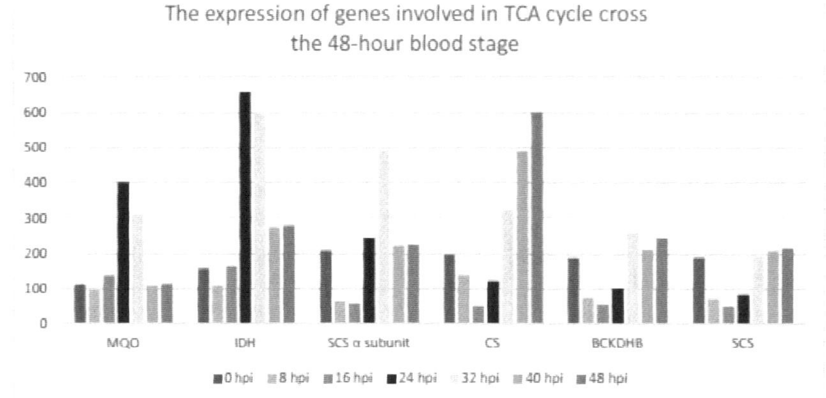

Figure 4. The expression of genes associated with TCA cycle across the 48-hour blood stage.

which is the precursor of thymidine nucleotides, Thymidylate Kinase (TMK), a ubiquitous enzyme that is important in the dTTP synthesis pathway for DNA synthesis, deoxyribose-phosphate aldolase (DERA), ribonucleotide reductases (RNR) and a putative gene of RNR, weresignificantly increased at 24 hpi, and most of them peaking at 32 hpi (Figure 5). Also, deoxyribonucleotide biosynthesis was concomitant with the induction of DNA replication and DNA repair machineries in the trophozoite stage.

invasion process. We also observed the induction of proteases involved in merozoite invasion, including three subtilisin-like proteases (SUB1-3) and three plasmepsins (PM VI, X and IX), which was consistent with previous studies [24]. In addition, twenty-two protein kinases were enriched in the schizont stage; they mediate specific protein phosphorylations which may be associated with merozoite release and reinvasion process.

The increased expression of 46 *var* genes, including 44 members of *var* family and two pseudogene for *var1CSA* and

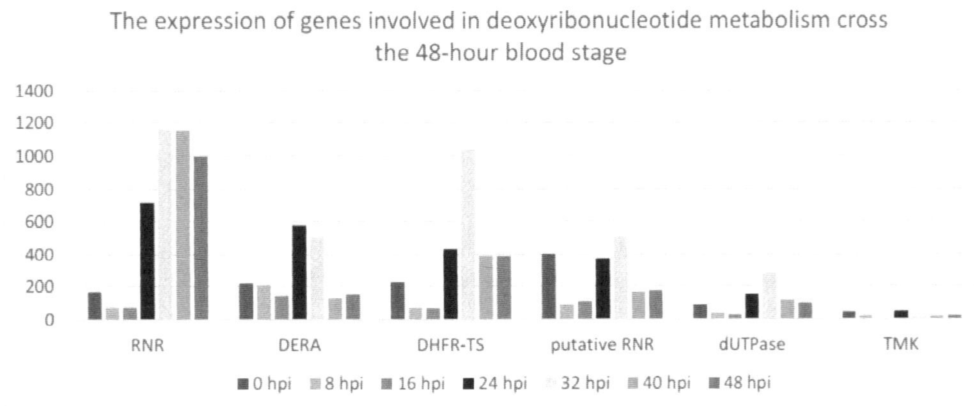

Figure 5. The expression of genes associated with deoxyribonucleotide metabolism across the 48-hour blood stage.

var, respectively, were detected in the schizont stage. They were concomitant with the reduced expression of the variant-silencing *SET* gene (*PfSETvs*), which is involved in broadly silencing *var* genes by trimethylating H3K36, suggesting that the simultaneous expression of *var* genes may contribute to the loss of inhibition by PfSETvs.

D. *The schizont stage*

A totally different set of genes were highly expressed in the schizont stage. The expression of genes associated with merozoite invasion was significantly increased at 40 hpi, including duffy binding-like merozoite surface protein (DBLMSP), six reticulocyte binding protein homologues (RH1, RH2b, RH3 pseudogene, RH4, RH5 and RH6 pseudogene) and RH5 interacting protein (RIPR); they may be involved in merozoite reorientation by binding receptors on the erythrocyte surface. In addition, other factors regulating the invasion of erythrocytes were also highly expressed in the schizont stage, including apical membrane antigen 1 (AMA1), required for merozoite junction formation by binding rhoptry neck protein 2 (RON2), the erythrocyte binding antigen-175 (EBA-175), which binds to glycophorin A on the surface of erythrocyte to release rhoptry proteins [2], erythrocyte binding antigen-181 (EBA181), which binds to a sialoglycoprotein on host cell and co-localizes with EBA-175 in the microneme organelles [23], cAMP-dependent protein kinase catalytic subunit (PKAc), which phosphorylates Ser610 on AMA1 [2], as well as eleven merozoite membrane proteins (MSP1-11) and three rhoptry proteins (RAP1-3). On the other hand, the components of actin-myosin motors, which may bring the merozoite into the erythrocyte, were activated in the schizont stage, including five myosin genes (*MyoA*, *MyoB*, *MyoD*, putative *MyoE* and putative *MyoF*), MyoA light chain (myosin A tail domain interacting protein), one actin-like protein (ALP5b) as well as five actin filament organization-associated proteins (actin-depolymerizing factor 2, coronin,formin 1, formin 2 and its putative protein). Therefore, our RNA-Seq analysis supported the hypothesis of multi-step merozoite

IV. CONCLUSION

Transcriptomic and ontological analyses of the IDC of *P. falciparum* revealed an orchestrated transcriptional machinery and a "just-in-time" mechanism for transcriptional regulation in the malaria parasite. For example, the reinvasion of erythrocytes is concomitant with the expression of genes associated with merozoite invasion; also, the deoxyribonucleotide metabolism, a trophozoite/early-schizont function [24], correlates well with the activation of enzymes that convert ribonucleotides into deoxyribonucleotides. The global analysis of the *P. falciparum* transcriptome defines the stage-specific biological processes by high resolution RNA-Seq. A future direction will be focused on delineating the cellular networks associated with transcriptional regulation, and characterizing temporal-specific antimalarial targets.

ACKNOWLEDGMENT

We thank PlasmoDB for providing access to malaria omic data. This work is supported by NIH grants GM100806, GM081068 and AI080579 to YW. This work received computational support from Computational System Biology Core, funded by the National Institute on Minority Health and Health Disparities (G12MD007591) from the National

Institutes of Health. The content is solely the responsibility of the authors and does not necessarily represent the official views of the National Institute of General Medical Sciences, National Institute of Allergy and Infectious Diseases, National Institute on Minority Health and Health Disparities, or the National Institutes of Health.

REFERENCES

[1] W. H. Organization, "World Malaria Report," 2015.

[2] L. H. Miller, H. C. Ackerman, X. Z. Su, and T. E. Wellems, "Malaria biology and disease pathogenesis: insights for new treatments," *Nat Med,* vol. 19, pp. 156-67, Feb 2013.

[3] T. D. Otto, D. Wilinski, S. Assefa, T. M. Keane, L. R. Sarry, U. Bohme, *et al.*, "New insights into the blood-stage transcriptome of Plasmodium falciparum using RNA-Seq," *Mol Microbiol,* vol. 76, pp. 12-24, Apr 2010.

[4] C. Aurrecoechea, J. Brestelli, B. P. Brunk, J. Dommer, S. Fischer, B. Gajria, *et al.*, "PlasmoDB: a functional genomic database for malaria parasites," *Nucleic Acids Res,* vol. 37, pp. D539-43, Jan 2009.

[5] M. D. Robinson, D. J. McCarthy, and G. K. Smyth, "edgeR: a Bioconductor package for differential expression analysis of digital gene expression data," *Bioinformatics,* vol. 26, pp. 139-40, Jan 1 2010.

[6] X. Zhou, H. Lindsay, and M. D. Robinson, "Robustly detecting differential expression in RNA sequencing data using observation weights," *Nucleic Acids Res,* vol. 42, p. e91, Jun 2014.

[7] M. Ashburner, C. A. Ball, J. A. Blake, D. Botstein, H. Butler, J. M. Cherry, *et al.*, "Gene ontology: tool for the unification of biology. The Gene Ontology Consortium," *Nat Genet,* vol. 25, pp. 25-9, May 2000.

[8] M. J. Blackman, "Malarial proteases and host cell egress: an 'emerging' cascade," *Cell Microbiol,* vol. 10, pp. 1925-34, Oct 2008.

[9] M. J. Meyers and D. E. Goldberg, "Recent advances in plasmepsin medicinal chemistry and implications for future antimalarial drug discovery efforts," *Curr Top Med Chem,* vol. 12, pp. 445-55, 2012.

[10] P. J. Rosenthal, "Falcipains and other cysteine proteases of malaria parasites," *Adv Exp Med Biol,* vol. 712, pp. 30-48, 2011.

[11] Y. Wu, X. Wang, X. Liu, and Y. Wang, "Data-mining approaches reveal hidden families of proteases in the genome of malaria parasite," *Genome Res,* vol. 13, pp. 601-16, Apr 2003.

[12] D. Jani, R. Nagarkatti, W. Beatty, R. Angel, C. Slebodnick, J. Andersen, *et al.*, "HDP-a novel heme detoxification protein from the malaria parasite," *PLoS Pathog,* vol. 4, p. e1000053, Apr 2008.

[13] H. J. Painter, T. L. Campbell, and M. Llinas, "The Apicomplexan AP2 family: integral factors regulating Plasmodium development," *Mol Biochem Parasitol,* vol. 176, pp. 1-7, Mar 2011.

[14] S. Iwanaga, I. Kaneko, T. Kato, and M. Yuda, "Identification of an AP2-family protein that is critical for malaria liver stage development," *PLoS One,* vol. 7, p. e47557, 2012.

[15] G. A. Josling, M. Petter, S. C. Oehring, A. P. Gupta, O. Dietz, D. W. Wilson, *et al.*, "A Plasmodium Falciparum Bromodomain Protein Regulates Invasion Gene Expression," *Cell Host Microbe,* vol. 17, pp. 741-51, Jun 10 2015.

[16] Q. Fan, L. An, and L. Cui, "Plasmodium falciparum histone acetyltransferase, a yeast GCN5 homologue involved in chromatin remodeling," *Eukaryot Cell,* vol. 3, pp. 264-76, Apr 2004.

[17] M. Yuda, S. Iwanaga, I. Kaneko, and T. Kato, "Global transcriptional repression: An initial and essential step for Plasmodium sexual development," *Proc Natl Acad Sci U S A,* vol. 112, pp. 12824-9, Oct 13 2015.

[18] B. Alberge, L. Gannoun-Zaki, C. Bascunana, C. Tran van Ba, H. Vial, and R. Cerdan, "Comparison of the cellular and biochemical properties of Plasmodium falciparum choline and ethanolamine kinases," *Biochem J,* vol. 425, pp. 149-58, Jan 1 2010.

[19] M. Fry, E. Webb, and M. Pudney, "Effect of mitochondrial inhibitors on adenosinetriphosphate levels in Plasmodium falciparum," *Comp Biochem Physiol B,* vol. 96, pp. 775-82, 1990.

[20] K. L. Olszewski and M. Llinas, "Central carbon metabolism of Plasmodium parasites," *Mol Biochem Parasitol,* vol. 175, pp. 95-103, Feb 2011.

[21] M. J. Gardner, N. Hall, E. Fung, O. White, M. Berriman, R. W. Hyman, *et al.*, "Genome sequence of the human malaria parasite Plasmodium falciparum," *Nature,* vol. 419, pp. 498-511, Oct 3 2002.

[22] L. W. Scheibel, S. H. Ashton, and W. Trager, "Plasmodium falciparum: microaerophilic requirements in human red blood cells," *Exp Parasitol,* vol. 47, pp. 410-8, Jun 1979.

[23] T. W. Gilberger, J. K. Thompson, T. Triglia, R. T. Good, M. T. Duraisingh, and A. F. Cowman, "A novel erythrocyte binding antigen-175 paralogue from Plasmodium falciparum defines a new trypsin-resistant receptor on human erythrocytes," *J Biol Chem,* vol. 278, pp. 14480-6, Apr 18 2003.

[24] Z. Bozdech, M. Llinas, B. L. Pulliam, E. D. Wong, J. Zhu, and J. L. DeRisi, "The transcriptome of the intraerythrocytic developmental cycle of Plasmodium falciparum," *PLoS Biol,* vol. 1, p. E5, Oct 2003.

Acceleration of Computational Fluid Dynamics Analysis by using Multiple GPUs

Hyungdo Lee[1], Bongjae Kim[2], Kyounghak Lee[3], Hyedong Jung[1]

[1]Embedded and Software Research Center, Korea Electronics Technology Institute, Korea
[2]Department of Computer Science and Engineering, Sun Moon University, Korea
[3]IACF Namseoul University, Korea
joytop88@keti.re.kr[1], bjkim@sunmoon.ac.kr[2], khlee@nsu.ac.kr[3], hudson@keti.re.kr[1]

Abstract - *GPU-based computing is widely used in various computing fields. In case of Computational Fluid Dynamics (CFD), there are computation intensive iterative solvers. Iterative solvers are bottlenecks of CFD. Recently, CFDs require high-accuracy and high-resolution than before. By above reason, the problem size of CFDs continues to grow and the performance of CFDs is also falling in terms of execution time. One of the solutions is to use GPU which support many cores than typical CPU. GPU can be used to accelerate the computation of CFDs like matrix multiplication. The improvement of GPU depends on how to use GPU due to the complexity of its architecture. In this paper, we propose a scheme to improve the performance of CFD applications based on multi-GPUs. In our approaches, we adjust GPU-based SpMV (Sparse Matrix Vector multiplication) and use multi-GPUs by considering characteristics of input matrix. We have changed the matrix multiplication method from scalar-based scheme to enhanced vector-based scheme. In addition, we used direct memory access (DMA) scheme among multi-GPUs to reduce the latency. Based on the performance evaluation result, the overall performance was improved 4.6 times when compare to previous CPU-based scheme.*

Keywords: Computational Fluid Dynamics, Multi-GPUs, CUDA

1 Introduction

Computational Fluid Dynamics (CFD) is a computer-based numerical analysis or simulation such as fluid flow and heat transfer [1]. Recently, CFDs requires high-accuracy and high-resolution. By above requirements, the size of problems is also increased continuously. For example, mesh structures for CFD analysis or simulation are getting fine-grained to obtain a more accurate result.

HPC (High Performance Computing) systems are essential and a good choice to deal with this problem. Because, HPC system supports massive computing power. In addition, GPUs can be applied to HPC systems to accelerate the computation. Typically, a GPU support more cores than typical CPU. For example, there are 2496 cores on each Tesla K20M GPU card. For these reasons, many studies have been performed in an effort to realize a high-performance computing environment based on GPU. Molecular dynamics [2][3], quantum chemistry [4], financial engineering [5][6], data mining [7] are some representative fields which use GPU-based HPC system.

In this paper, we focus on CFD application. Typically, there are many iterative solvers in CFD applications. iterative solvers is a dominant part of CFD application in terms of execution time. Iterative solvers are bottleneck of CFD simulation. Therefore, iterative solver is key point to increase the performance of CFDs. An iterative solver is mainly consisted of SpMV (Sparse Matrix Vector multiplication). GPU is one of the best solutions to accelerate SpMV computation. However, GPU architecture is very complex and it is hard to obtain relatively good performance based on GPU. In this paper, we propose some scheme and approaches to increase the performance in terms of execution time and latency. To increase the performance of CFDs, we adjust GPU-based SpMV method and use multi-GPUs by considering characteristics of input matrix. In our SpMV scheme, a warp (32 GPU threads) are assigned to multiple rows of a sparse matrix to calculate matrix vector multiplication. Because, a warp is a scheduling unit of GPU. By using this manner, we can minimize the memory access of GPUs when doing SpMV. In addition, we use direct memory access scheme to reduce the data transfer latency among multi-GPUs. Our enhanced SpMV scheme is applied to BiCGStab and CG solver. BiCGStab and CG solver are two representative iterative solver algorithms which are widely used in CFDs. Based on the performance evaluation result, the overall performance was improved 4.6 times when compare to previous CPU-based approach.

The rest of this paper as follows. In section 2, some related works are discussed. In section 3, we will explain our approach to improve the performance based on the computing environment with multi-GPUs. Performance evaluation. Finally, we conclude this paper with future works in section 5.

2 Related Works

2.1 3D Coronary Artery Blood Flow Dynamics

CFD is commonly used in scientific and engineering fields to investigate fluid flow and its interactions in a particular domain. In order to applying our multi-GPU accelerating scheme, we use 3D coronary blood fluid-dynamics simulation application which investigates flow of blood and its interactions to diagnose cardiovascular disease. This application is constructed using 3D unsteady Navier-Stokes equations which describe how the velocity, pressure of a moving fluid is related. And our 3D unsteady Navier-Stokes equations use Finite Element Method (FEM). FEM subdivides a large problem into smaller, simpler, parts, called finite element. So it can make discretized domain from physical space. Many studies have shown that finite element methods can be successfully applied to the analysis of the unsteady Navier-Stokes equations [8]. and this application use Uzawa iteration to solve the Navier-Stokes equation by using FEM. Uzawa Iteration consist of outer iteration to update the pressure and an elliptic inner iteration for velocity. In computing 3D unsteady Navier-Stokes equations, we should consider time as a fourth coordinate direction. As space coordinates are discretized, time must be discretized. And explicit method is used for time integration. It is computed before executing Uzawa iteration [9]. In this way, the 3D coronary blood fluid-dynamics is analyzed. Uzawa iteration and explicit method has linear system problems. So Conjugate Gradient (CG) and Bi-Conjugate Gradient Stabilized (BiCGStab) solvers can solve the problems.

2.2 Preconditioned Iterative Solver

The size of modeling of engineering, physics and economics is increasing. So solving the large-scale linear systems is essential. For solving the large-scale linear systems, direct method is no effective. Because it has a heavy memory load and the computing overhead. From the past, iterative solver like CG, GMERS, BiCGStab were designed to overcome direct method problems in terms of performance. And preconditioning techniques also were designed to improve accuracy and performance of iterative solver [10]. As other studies to improve the iterative solver in progress, many solvers were proposed like Bi-Conjugates Gradient (BiCG), Conjugate Gradients-Squared (CG-S), Bi-Conjugate Gradients Stabilized (BiCGStab) solver [11]. CG-S is a variant of the BiCG solver. But, it has been observed that CG-S may lead to a rather irregular convergence behaviour, so that in some cases rounding errors can even result in severe cancellation effects in the solution. So, another variant of BiCG or BiCGStab which is more stabilized and efficient. In this paper, explicit method uses BiCGStab solver and Uzawa iteration use CG solver to solve linear system at unsteady Navier-Stokes equation.

When solving a linear equation of the form $A \times x = b$ for x, where matrix A is large and b is a vector, time of obtaining x should be a long time. Because computational complexity of matrix inversion. In this case, the iterative solver is used. Furthermore, after substitute the form $A \times x = b$ to $r_i = b - A \times x$, the iterative solver repeat until r_i becomes sufficiently small. This allows to obtain the similar approximations as x. and preconditioner make it fast and accurate.

In recent years, the iterative solver has been applied to HPC with GPU [12]. also the preconditioner has been applied [13]. Furthermore, study of improving CG solver using texture memory and shader function of GPU are in progress [14].

2.3 Compute Unified Device Architecture

Single core of CPU has some limitation in terms of it performance like computing power. So multi-core architecture like dual or octa cores has been applied to CPU architecture to improve the performance. However, there is also a limit that increasing the number of CPU cores in a single chip. In order to overcome this problem and achieve HPC, many-core architecture with GPU can used to general purposes. The reason why GPU has many cores is that common graphics applications handle large 3D rendering and multi-textures and these require huge computing power. Formerly, general purpose programming in the GPU-based computing environment was not possible. However, the needs of HPC equipped with a lot of GPU devices has been increasing continuously. So GPU vendors are developing GPU platform including GPU programming language and programming tools. The latest released GPU's performance has the minimum 1 TFLOPS/s of computation performance and 160GB/s of off-chip memory bandwidth. By above reasons, many GPU-based HPC system are widely used for various computing fields. There are two major GPU platforms. First one is OpenCL(Open Computing Language) which is used universally today, Second one is CUDA (Compute Unified Device Architecture) which can be used over only NVIDA GPUs or architecture. In this paper, we use NVIDIA GPUs and CUDA programming model to improve the performance of CFD applications.

3 CFD Acceleration by using Multi-GPUs

3.1 Basic Iterative Solver based on CUDA Programming Model

First of all, in this sub-section, we explain an approach that accelerates BiCGStab and CG solver by using single GPU. GPU-based computing acceleration is a kind of data parallelization. Therefore, if there are data dependency in algorithms, it is hard to parallelize the algorithm with GPU. In case of BiCGStab and CG solver are mainly consisted of SpMV operation, addition and subtraction between vectors, and inner product operation between vectors. These operations are very suitable to apply GPU-based computing acceleration because these operations do not have any data dependency. In addition, some preconditioner can be used to converge rapidly for BiCGStab and CG solver. There are many preconditioners such as diagonal, incomplete factorization, approximate inverse preconditioner. In our scheme, we choose the Jacobi preconditioner because it is very suitable to parallelize its operation. We use CUDA programming model like Figure 1 to parallelize those operations and those operations can be executed on the GPU in parallel.

$$\text{for } i = 1, 2, 3, \ldots, n \qquad gId = blockIdx * blockDim + threadIdx$$
$$v_{(i)} = a_{(i)} + b_{(i)} \qquad v_{gId} = a_{(gId)} + b_{(gid)}$$

Figure 1. An Example of Vector Addition Code (Left Side: Sequential(CPU), Right Size: Parallel(GPU))

Figure 1 shows an example of vector addition code. Left side code is a sequential version for CPU. Right side code is a parallel version which is followed CUDA programming model for GPU. In case of the sequential version, as shown in Figure 1, the addition process sequentially proceeds *for* statement from first to last elements of the vector to calculate vector *v* [15]. In case for GPU, on the other hand, the addition process creates many threads as much as the size of vector, *n* by using CUDA C like *blockIdx*, *blockDim* and *threadIdx*. And the thread of the size of the vector is assigned to concurrently *gId*. A thread to process the addition take each element from two vectors. After proceeding the addition, a thread stores the added value in the vector *v*. This progress can be executed concurrently at the GPU.

```
1.   gId = blockIdx*blockDim + threadIdx
2.   for (k = rowPtr[gId]; k < rowPrt[gId+1]; k++)
3.       r[gId] += val[k] * v[culm[k]]
```

Figure 2. GPU-based SpMV Code Using CSR Format

Figure 2 is an accelerated SpMV code using CUDA programming model. Similar to Figure 1, SpMV operation can be accelerated with GPU and CUDA programming model. A matrix used in the our SpMV is a form of sparse matrix. It needs to be compressed to use memory efficiently and compute its related operation fast. There are many methods to compress a sparse matrix to a compressed form. For examples, there are Coordinate (COO), ELLPACK (ELL), Hybrid (HYD) and CSR (Compressed Storage Row) format. Those format is classified according to the nature of the sparse matrix. In short, the proper compress algorithm is different depending on the characteristic of a sparse matrix. DIA, ELL, CSR, HYB, COO format is sorted by usage of the nature of the matrix. The order is from structured matrix to unstructured matrix. For example, DIA format is very suitable for structure matrix form. COO format is very suitable for unstructured matrix form. The nature of sparse matrix of our blood fluid dynamics is middle of structured and unstructured matrix. Therefore, we use CSR matrix format when compress an original sparse matrix.

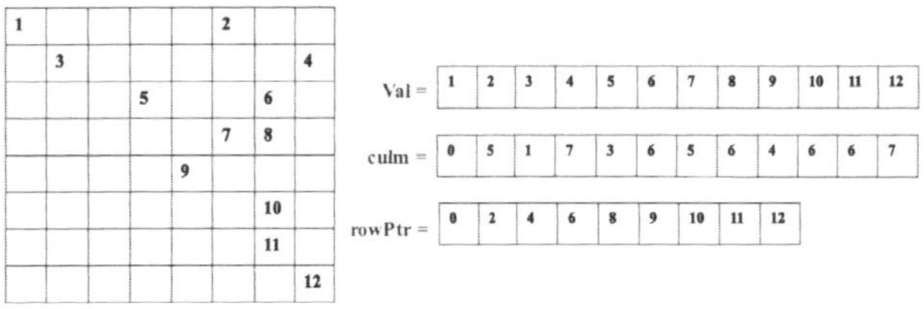

Figure 3. An Example of Sparse Matrix and its CSR Format

Figure 3 is an example of sparse matrix and its CSR format. As shown in Figure 3, the size of sparse matrix is 10 × 10. First, non-zero values of sparse matrix sequentially store in the *val*. Second, *val*'s column position at the sparse matrix is stored in the *culm* per each non-zero value. At last, *rowPtr* stores the starting index of each row *in Val*.

3.2 Advanced Iterative Solver based on Memory Coalescing

In our basic iterative solver, we just use a scalar-based SpMV. One GPU thread is assigned per row of sparse matrix in case of the scalar-based SpMV. Typically, a warp is a scheduling unit of GPU. A warp is consisted of 32 GPU threads. In basic iterative solver, a warp cannot access to contiguous memory space in which non-zero elements are stored. By above reason, scalar-based SpMV could not utilize memory coalescing when access to memory. If we use memory coalescing, we can reduce the number of memory access operation for a warp to only one memory access operation.

In the advanced iterative solver, we used enhanced SpMV method to improve the performance more. In the advance iterative solver, a warp (32 GPU threads) are assigned to multiple rows of a sparse matrix to calculate matrix vector related operations. A warp can access to contiguous memory space by this manner. Therefore, multiple memory access operations can be minimized. Figure 4 shows an example of GPU thread assignment comparison between scalar-based SpMV and enhaced SpMV.

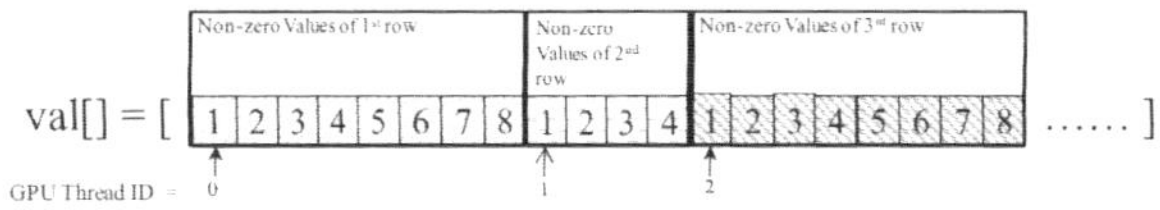

(a) GPU Thread Assignment of Scalar-based SpMV (Basic Iterative Solver)

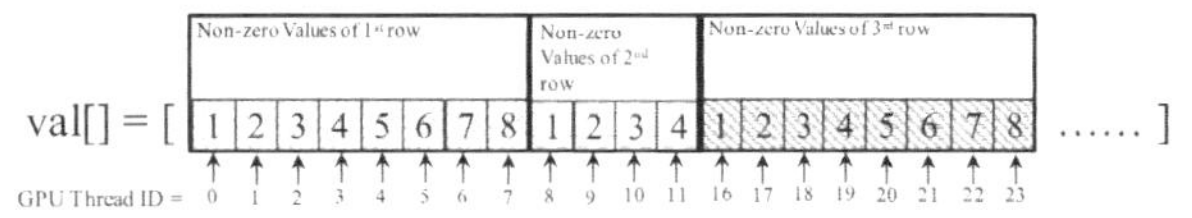

(b) GPU Thread Assignment of Enhanced SpMV (Advance Iterative Solver)

Figure 4. An Example of GPU Thread Assignment Comparison Between Scalar-based SpMV(Basic Iterative Mode) and Enhanced SpMV (Advanced Iterative Solver)

3.3 Using Multi-GPUs with Domain Decomposition and MPI Programming Model

If we use domain decomposition scheme, we can obtain further acceleration with multi-GPUs. In our scheme, we device main domain into multiple sub-domains. Each sub-domain for CFDs is assigned to one GPU to simulate or analyze the result [16]. Figure 5 show the concept of using multi-GPUs with domain decomposition.

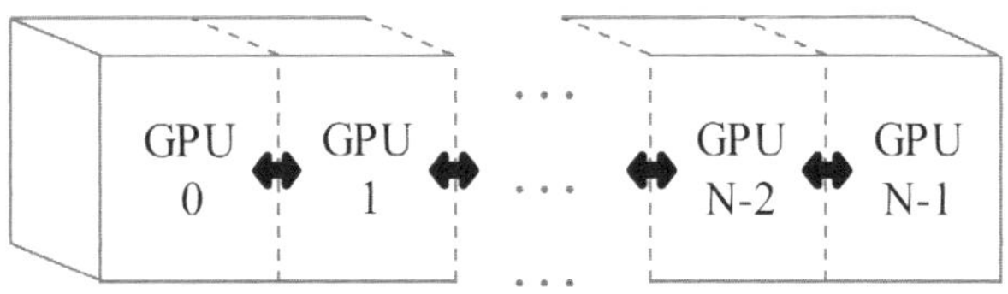

Figure 5. A Concept of Using Multi-GPUs with Domain Decomposition

If we use domain decomposition scheme, data exchanged between two adjacent domains or GPUs because intermediate result affect each adjacent domain. In our approaches, we use MPI (Message Passing Interface) programming model to communicate and exchange the intermediate result between two adjacent domains or GPUs.

In this situation, there are 2 step data copy operations. For example, first step is one GPU memory to main memory, and second step copy is required from main memory to another GPU memory. Data communication overhead is considerable. This communication is also bottleneck of GPU-based computation. To reduce the data communication overhead, we use DMA (Direct Memory Access) scheme between two GPU memory. By using DMA, we can communicate directly among multi-GPUs. In short, we can improve the performance more by using DMA.

4 Performance Evaluation

4.1 Evaluation Environment

Table 1 shows the computing environment used in the performance evaluation. GPU and CUDA libraries are necessary must be installed in the system to apply CUDA-based programming model onto preconditioned iterative solver for CFD simulation or analysis. As shown in Table 1, CUDA 7.5 was installed in the computing system. GeForce Titan Black and Tesla K20 are equipped. In case of MPI library, MVAPICH2 2.2b was used to evaluate the performance.

Table 1. Computing Environment

Features	Descriptions
CPU	2 × Intel Xeon CPU E5-2650 2.6 GHz v2
Memory	64 GB
OS	CentOS 7.2
Kernel Version	3.10.0
CUDA Version	CUDA 7.5
OFED Version	MLNX_OFED_LINUX 3.2
MVAPICH2 Version	MVAPICH2-2.2b
GPU	GeForce Titan Black Tesla K20
InfiniBand Host Channel Adapter	Mellanox ConnectX-3 VPI Adapter Dual-Port QSFP, FDR IB(56 Gb/s)

4.2 Evaluation Results

4.2.1 Acceleration of Basic Iterative Solver based on CUDA Programming Model

Table 2 shows the result of acceleration of basic iterative solver per one iteration. The row size of sparse matrix for CG is 219,725 and non-zero elements are 2,710,418. In case of the BiCGStab, the row size is 219.725 and its non-zero elements are 5,201,235. As shown in Table 2, the maximum speed up is 7.349 by GPU-based acceleration. Overall, GPU-based acceleration is better than CPU-based solver. Titan Black is better than Tesla K20 because the base clock speeds of Titan Black and Tesla K20 are 889 MHz and 706 MHz, respectively.

Table 2. Basic Iterative Solver Results

Features		CPU (s)	GPU (s)	Speed Up
CG	Tesla K20	0.716	0.195	3.672
	Titan		0.115	6.226

	Black			
BiCGStab	Tesla K20	0.801	0.167	4.811
	Titan Black		0.109	7.349

4.2.2 Acceleration of Advanced Iterative Solver based on CUDA Programming Model

Table 3 shows the result of acceleration of advanced iterative solver per one iteration. As shown in Table 3, the maximum speed up is 18.372 seconds. Maximum speed up is increased from 7.347 seconds to 18.372 seconds due to memory coalescing. Similar to Table 2, Titan Black is better than Tesla K20 in terms of the execution time. Based on the Table 2 and Table 3, we can improve the performance of CG and BiCGStab solver. As the result, we can reduce the overall execution time of CFD simulation or analysis application.

Table 3. Advanced Iterative Solver Results

Features		CPU (s)	GPU (s)	Speed Up
CG	Tesla K20	0.716	0.107	6.692
	Titan Black		0.073	9.808
BiCGStab	Tesla K20	0.801	0.062	12.919
	Titan Black		0.044	18.372

4.2.3 Intra-node Data Transfer Latency between GPUs

Table 4 shows the result of intra-node data transfer latency between two GPUs. In the performance evaluation, we used Mellanox OFED 3.2, MVAPICH2 2.2b, and NVIDIA GeForce Titan Black. As shows in Table 4, In case of 32 KB data transfer, DMA is about 1.2 times better than No DMA. Similarly, in case of 64 KB data transfer, DMA is about 5.9 times faster than No DMA. As we explained in Section 3, The overhead of data exchange or transfer is considerable between two intra GPUs. Therefore, if we use DMA, we can reduce the data transfer delay effectively.

Table 4. Intra-node Data Transfer Latency Between GPUs

Size (Byte)	No DMA (us)	DMA (us)
32 K	75.8	59.0
64 K	124.9	59.4
128 K	197.9	59.4
256 K	366.9	61.5

Acceleration of Blood Fluid Dynamics by using Multi-GPUs and Advanced Iterative Solver

To evaluate the performance of our scheme, we used blood fluid simulation application. Advance iterative solver and DMA scheme were applied to our blood fluid simulation application. BiCGStab and CG solver were used.

Figure 6 shows the result of total execution time of blood fluid dynamics. In the performance evaluation, we used GeForce Titan Black GPU because Titan Black showed better performance than K20 GPU. Simulation areas are divided into 4 areas. Each area is assign to one CPU core or one GPU device. As shown in Figure 6. We can reduce the total execution time of blood fluid dynamics by using multi-GPUs. The total execution time was reduced from 7113 seconds to 1527 seconds. In short, the performance was improved about 4.6 times in terms of speed up.

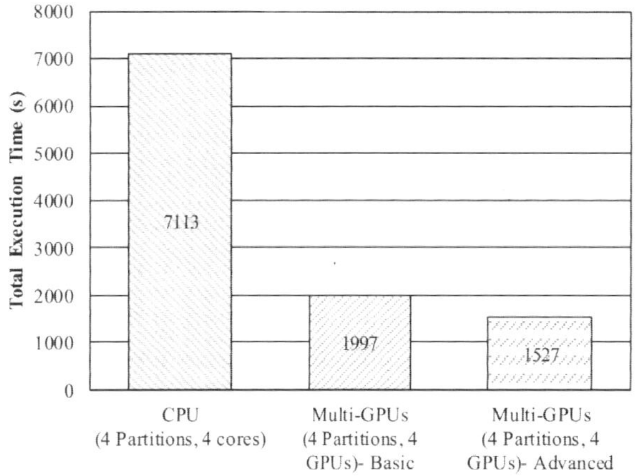

Figure 6. Total Execution Time of Blood Fluid Dynamics

5 Conclusions and Future Works

In this paper, we proposed a scheme to improve the performance of CFD application based on multi-GPUs. Our scheme used enhanced vector-based SpMV method with domain decomposition in order to reduce the SpMV time. SpMV is a dominant part of CFD applications in terms of execution time. We reduced the SpMV time by using multi-GPUs. In addition, DMA scheme are also used to reduce the latency among multi-GPUs. As the result, overall execution time decreased efficiently. Based on the performance evaluation results, the performance was improved 4.6 times when compared to the previous scheme which uses 4 CPU cores. In the future works, we will apply GPU-based computing to other computation-sensitive fields.

6 Acknowledgement

This work was supported by Institute for Information & communications Technology Promotion(IITP) grant funded by the Korea government(MSIP) (No.R0101-15-0171, Development of Multi-modality Imaging and 3D Simulation-

Based Integrative Diagnosis-Treatment Support Software System for Cardiovascular Diseases)

7 References

[1] Ferziger, Joel H., and Milovan Peric. Computational methods for fluid dynamics. Springer Science & Business Media, 2012.

[2] Anderson, Joshua A., Chris D. Lorenz, and Alex Travesset. "General purpose molecular dynamics simulations fully implemented on graphics processing units." Journal of Computational Physics 227.10 (2008): 5342-5359.

[3] Anderson, Joshua A., Chris D. Lorenz, and Alex Travesset. "General purpose molecular dynamics simulations fully implemented on graphics processing units." Journal of Computational Physics 227.10 (2008): 5342-5359.

[4] Olivares-Amaya, Roberto, et al. "Accelerating correlated quantum chemistry calculations using graphical processing units and a mixed precision matrix multiplication library." Journal of chemical theory and computation 6.1 (2009): 135-144.

[5] Fatone, Lorella, et al. "Parallel option pricing on GPU: barrier options and realized variance options." The Journal of Supercomputing 62.3 (2012): 1480-1501.

[6] Surkov, Vladimir. "Parallel option pricing with Fourier space time-stepping method on graphics processing units." Parallel Computing 36.7 (2010): 372-380.

[7] Jian, Liheng, et al. "Parallel data mining techniques on graphics processing unit with compute unified device architecture (CUDA)." The Journal of Supercomputing 64.3 (2013): 942-967.

[8] Taylor, Cedric, and P. Hood. "A numerical solution of the Navier-Stokes equations using the finite element technique." Computers & Fluids 1.1 (1973): 73-100.

[9] Ferziger, Joel H., and Milovan Peric. Computational methods for fluid dynamics. Springer Science & Business Media, 2012.

[10] Elman, Howard C. Iterative methods for large, sparse, nonsymmetric systems of linear equations. Diss. Yale University, 1982.

[11] Van der Vorst, Henk A. "Bi-CGSTAB: A fast and smoothly converging variant of Bi-CG for the solution of nonsymmetric linear systems." SIAM Journal on scientific and Statistical Computing 13.2 (1992): 631-644.

[12] Oancea, Bogdan, Tudorel Andrei, and Andreea Iluzia Iacob. "CUDA based iterative methods for linear systems." Computer Science 1 (2012): 228-232.

[13] Dehnavi, Maryam Mehri, et al. "Parallel sparse approximate inverse preconditioning on graphic processing units." Parallel and Distributed Systems, IEEE Transactions on 24.9 (2013): 1852-1862.

[14] Bolz, Jeff, et al. "Sparse matrix solvers on the GPU: conjugate gradients and multigrid." ACM Transactions on Graphics (TOG). Vol. 22. No. 3. ACM, 2003.

[15] Cormie-Bowins, Elise. "A comparison of sequential and GPU implementations of iterative methods to compute reachability probabilities." arXiv preprint arXiv:1210.6412 (2012).

[16] Jacobsen, Dana A., Julien C. Thibault, and Inanc Senocak. "An MPI-CUDA implementation for massively parallel incompressible flow computations on multi-GPU clusters." 48th AIAA aerospace sciences meeting and exhibit. Vol. 16. 2010.

RNAseq Analysis of C. elegans infected by Orsay virus suggests evidence of a new RNAi or stress related pathway

Jessica Ngo [1], Jahanshah Ashkani [2], Frederic Pio[1]
[1] Mol. Biol. & Biochem. Department, Simon Fraser University, Burnaby, B.C., Canada
[2] Scientific Computing Research Unit, University of Cape Town, Cape Town, South Africa

Abstract - *The C. elegans genome is currently the best annotated eukaryotic genome. As a result, the identification of novel genes and their functional annotation is challenging. The recent discovery of the Orsay virus in C. elegans offers an attractive possibility. Effectively, in contrast to the exogenous RNAi pathway, the antiviral RNAi response is not systemic and cannot be passed between generations, as shown for plants and other eukaryotes. These results suggest that the two antiviral RNAi pathways between C. elegans and plants are different and the genes involved remain to be discovered. In this study, we performed a meta-analysis using BIOCONDUCTOR of RNAseq data deposited in GEO express at NCBI and found that the top 17 genes that are differentially expressed during infection based on p value are novel uncharacterized genes. Further attempts to annotate these genes suggest their involvement in a new RNAi or stress-related response pathway.*

1 Introduction

Since its discovery in C. elegans, RNA interference (RNAi) has proven to be essential during development and in diseases and a valuable scientific tool in many organisms by silencing gene expression. Exogenous RNAi spreads throughout the organism between cells and can be passed between generations; however, there has been controversy as to the endogenous role(s) that the RNAi pathway plays. One endogenous role for which spreading both within the infected organism and between generations would be advantageous is a role in viral defense. In plants, antiviral RNAi is systemic and the spread of RNAi between cells provides protection against subsequent viral infection [1]. Recent studies, on the Nodavirus Orsay able to infect C. elegans [2], found that in contrast to the exogenous RNAi pathway, the antiviral RNAi response targeted against this virus does not spread systemically throughout the organism and cannot be passed between generations. These results suggest that there are differences between the two RNAi pathways between C. elegans and plants that remain to be discovered [1]. In this study, we analyzed RNAseq data deposited in the GEO database at NCBI to identify genes responsible for these differences [3].

2 Method

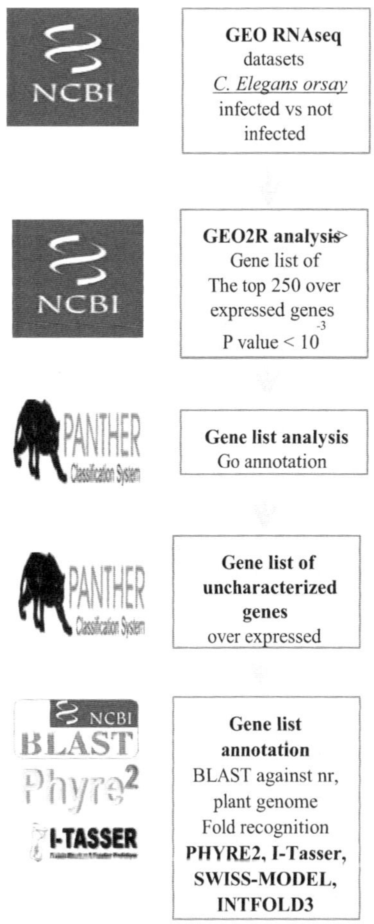

In the pipeline presented in this study, GEO datasets from C. elegans infected by Orsay virus (GSE41056) were processed by the GEO2R RNAseq analysis package that uses the R package from Bioconductor called LIMMA with the libraries Biobase and GEOquery [4][5][6]. In brief, (i) a series of GEO expression datasets specified by their accession numbers were queried and entered (ii) two sample groups were defined as infected and non-infected (iii) RNAseq samples were then assigned to each group (iv) the test was then run and genes that were differentially expressed were ranked according to their p-value to select for the top 250 genes. As a result, a list of ordered genes differentially expressed was obtained with an adjusted p value ranging from $9.82*E^{-11}$ to $9.88*E^{-04}$. In our analysis, the p value was adjusted by correcting for error introduced by multi-testing according to the method of Benjamini et al [4][7]. In an attempt to annotate these genes and understand in which pathway those differentially expressed genes were involved we further performed a gene set enrichment analysis using the panther package call gene list analysis. This package provided from a list of genes, their Gene Ontology (GO) annotation and determined to what extent a particular gene function, biological process or pathway was enriched through the calculation of an enrichment factor [8]. The 250 differentially expressed genes were used as input for annotation.

To further annotate the function of these uncharacterized genes we predicted their structure using different fold recognition servers then infer their functions from the 3D-model obtained. Amino acid sequences were initially submitted to Phyre and i-Tasser servers [9][10]. Since the predictions were not giving significant results, we performed a more exhaustive search for folds, using additional methods from SWISS-MODEL, and the INTFOLD3 server [11][12]. The structures from the three different programs were then structurally aligned and compared to look for a consensus in the predictions across the different programs.

Additionally, we determined into what common biological processes these genes were involved in. Each gene name was used as keyword to query and identify the GEO data sets (from GEO express database at NCBI) where their expression was changed [3]. A filter selecting for differential expression of the gene was used. This process was repeated without the filter for the genes that came up with "no results found" when the filter was previously on, to check whether the gene was constitutively expressed, or whether the gene simply did not exist within the data set. The abstracts of the GEO data set of the 17 individual searches were combined together. A text mining word cloud approach was used to identify common words between the abstracts that may indicate their involvement in common biological processes

The STRING database of protein-protein interactions was then used to build protein-protein interaction networks of these genes to explore further into their functions. We determined how they may interact with other genes of known function and in which process, and using the guilt by association principle infer their function [13].

3 Results

3.1 Go annotation

GO annotation using panther (Table 1) revealed that most of the genes differentially expressed are the genes involved in catalytic activity and nucleic-acid binding [8]. However, review of this data using gene set enrichment analysis revealed that the top genes that had a significant p-value were genes involved in DNA repair, response to stress, the cell cycle and catabolic processes.

Table 1
Enrichment factors of the GO categories for the 250 genes differentially expressed after orsay virus infection. P value adjusted by bonferroni correction of multiple testing [14].

PANTHER GO	#	expected	∇ Enrichment	+/-	P value
DNA repair	7	1.09	6.40	+	2.58E-02
stress	17	3.52	4.83	+	2.48E-05
catabolism	13	2.74	4.74	+	9.44E-04
cell cycle	24	5.89	4.07	+	1.45E-06
catalytic activity	15	3.86	3.88	+	1.90E-03
regulation	15	3.88	3.86	+	2.02E-03
translation	13	3.45	3.77	+	1.00E-02
nitrogen compound metabolism	20	5.61	3.56	+	2.23E-04
phosphorylation	14	4.41	3.18	+	3.13E-02

Table 2
The 17 top ranking Gene differentially expressed based on the lowest p value have no known function [15].

Gene symbol	P value
B0507.8	9.82E-11
F26F2.4	1.26E-09
F26F2.5	3.57E-09
B0507.10	4.69E-09
CELE_T26F2.3	7.51E-09
CELE_C43D7.4	2.01E-08
CELE_C17H1.6	4.54E-08
CELE_C17H1.7	6.22E-08
CELE_Y75B8A.39	1.81E-07
F26F2.2	3.79E-07
CELE_C43D7.7	4.07E-07
F26F2.3	4.08E-07
F26F2.1	4.81E-07
C49C8.2	1.73E-06
CELE_B0284.4	1.90E-06
F42C5.3	7.59E-06
sdz-6	1.21E-05

Quite interestingly, we also noticed that many of the unclassified genes were overrepresented among the 250 genes (p value near 0), but with no known enrichment factors since these genes were annotated as unclassified. We went back further to the list of differentially expressed genes obtained from GEO2R to determine how significant their p value was in term of differential expression. Our results showed that the top 17 genes among the 250 differentially expressed genes after the Orsay virus infection currently do not have any known function (table 2) [15].

3.2 Annotation of the uncharacterized genes

To further annotate the function of the 17 genes that were the most differentially expressed and had no function, we performed a protein blast analysis in different databases [16]. The nr database, which contained all known protein sequences and also the plants database of protein sequences were used. If significant hits between plants and C. elegans were found, it could be that we had identified functional insights that were specific of the RNAi response common between plant and C. elegans. Remarkably, using BLAST we were not able to detect any significant hits on either of the databases mentioned. This suggests that these genes are very specific of C. elegans and maybe contribute to the specific antiviral response that is not systemic and not transgenerational in this organism. Then, to further annotate these genes using a more sensitive method, we performed comparative modeling using PHYRE2 and i-tasser [9][10]. In this procedure, a sequence alignment was searched between the C. elegans protein sequence of an unknown function and a protein sequence of known structure. A three dimensional structure of the sequence was then built and a statistical score was given for the sequence alignment and the quality of the three dimensional model. Finally, a functional inference on the uncharacterized C. elegans sequence could then be determined based on what we knew about the function of the protein sequence of known structure from which the model was built.

The results of the structural prediction obtained by PHYRE2 and i-tasser (table 3) showed that the coverage of the C. elegans protein sequences by the model built from the alignment was very low (mostly around (18 to 37%)) [9][10]. The confidence score of the structural prediction was also very low or not significant (NS). In most instances, the confidence score was around 41% which is also weak. For the predictions with a confidence score of 95% and up, they were always predictions of α-helicoidal proteins and in a region covering only a small part of the protein sequence of C. elegans. These predictions may indicate protein or nucleic acid binding through α-helicoidal regions, but these genes do not seems to have any real functional orthologues in any database. This data strongly shows that we have identified 17 novel genes that are really specific and important to the C. elegans RNAi response to viral infection.

When the genes were modeled again across three different servers with updated templates, in a second attempt to annotate the genes, it was clear that for most of these genes, models created would converge towards the secondary structure of an α-helix but not to a fold. It further supports that the fold of these proteins is either unknown or not represented in the structural databases used by these fold recognition methods.

Table 3
Initial structural prediction and annotation using PHYRE2 and i-tasser [9][10]. Confidence of the prediction and coverage of the C. elegans protein sequence are given in percentages.

Gene symbol	PHYRE Confidence (coverage)%	Annotation Based on template function
B0507.8	NS	
F26F2.4	NS	
F26F2.5	NS	
B0507.10	96.1(27)	Beta-myosin (α-helix)
CELE_T26F2.3	NS	
CELE_C43D7.4	NS	
CELE_C17H1.6	95.3(37)	Fibrinogen, (α-helix)
CELE_C17H1.7	96.1(27)	Same as C17H1.6)
CELE_Y75B8A.3	94(23)	Transcription regulator (mafg)
F26F2.2	41.3(18)	Zinc finger
CELE_C43D7.7	41(18)	Csm1 replication
F26F2.3	41.3(18)	Oxidoreductase (pnpc)
F26F2.1	NS	
C49C8.2	NS	
CELE_B0284.4	95.8(32)	Fibrinogen (α-helix)
F42C5.3	NS	
sdz-6	NS	

However, for the gene CELE_T26F2.3, upon the second attempt, we observed that the models created across the three programs converged into a single fold, even though a different template was used for the SWISS-MODEL server (pdb:5bto) [11]. Further investigation into gene function using WormBase revealed that this gene has been annotated as a vertebrate homologue of a de-capping exonuclease called DXO/Dom3Z which agrees with the template used for both the Phyre and INTFOLD3 predictions [17].

3.3 GEOexpress queries

Since our analysis has not been able to determine the function of most of these genes. It may be that the common biological process in which they are involved is also unknown. We decided to query GEO datasets that show differential gene expression of the 17 uncharacterized genes [3]. Their gene expression changes obtained in similar conditions may shed light on the type of biological process they may be involved in. A Word cloud analysis of all the abstracts referring to the GEO data sets combined from each gene query indicated that the genes expression changes were occurring during stress, metal toxicity responses and development. Many instances of the transcriptional regulator E2F was also obtained. Since it is an important regulator of the immune and stress response. Involvement of the E2F transcription factor may suggest that these genes have E2F binding site, and are a target of an existing or new unclassified E2F transcription factor family member (Table 4). From these data, we propose that these genes may be involved in a new stress related response pathway.

Table 4
Text mining of the GEO data sets abstracts showing differential expression of the 17 uncharacterized genes using word cloud. It should be noted that words that were important, but repeated extensively due to the subject of their paper, were removed. These words include: heme, HRG, LIN, cell, cadmium, pocket and transcription.

3.4 STRING Protein-protein Interactions

Additionally, we expanded upon the GEOexpress results by using the STRING protein-protein interaction analysis and discovered that 7 of these genes are shared amongst other species in the genus Caenorhabtitus. These genes may actually be a part of a novel pathway in response to stress that is conserved in, and specific to, this genus.

To gain insight into the function of these genes we further made an effort to increase the protein-protein interaction network of these uncharacterized genes by searching for some interacting partners or neighbors that have known function. In Table 5 the brown circle surrounds the 17 uncharacterized genes in the center of the network. The edges that are colored based on the interaction's evidence and that connect proteins, show only co-expression linkages for the 17 genes under study. None of the associated gene clusters are highlighted, as they have no assigned biological process in PANTHER [8].

Table 5
Protein-Protein interaction Network of the linkage between the 17 uncharacterized genes and their associated partners in STRING database [13]

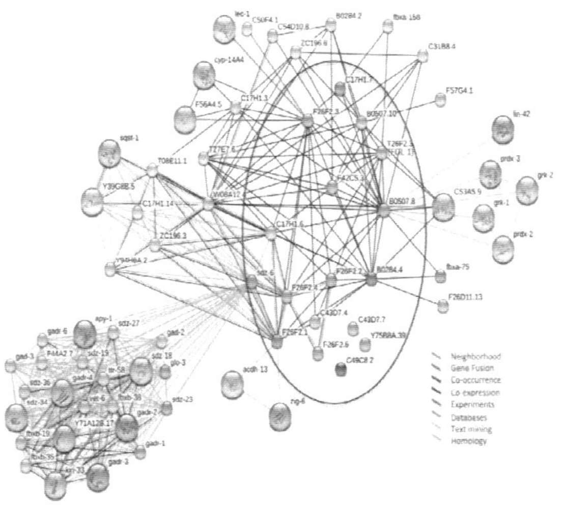

4 Conclusion

In this study we have identified 17 novel genes that are uncharacterized, and maybe specific of the antiviral RNAi response of C. elegans. Through the use of structure prediction, we were able to annotate only one of these genes (T26F2.3) as a Dom3Z homologue. Further analysis suggest the importance of these genes in a stress related pathway that maybe specific of C. elegans. Additional functional studies are needed to unravel this(ese) pathway(s) that are likely to be new.

5 References

[1] Ashe A, et al. "RNA Interference against Orsay Virus Is neither Systemic nor Transgenerational in Caenorhabditis elegans"; J Virol, 89, 23, Dec 2015.

[2] Félix M et al. "Natural and experimental infection of Caenorhabditis nematodes by novel viruses related to nodaviruses"; PLoS Biol, 9, 1, Jan 2011.

[3] Barrett T et al. "NCBI GEO: archive for functional genomics data sets – update"; Nucleic Acids Reseach, 41(Database issue), D991-D995, Jan 2013

[4] Benjamini Y. et al. "Controlling the false discovery rate: a practical and powerful approach to multiple testing"; Journal of the Royal Statistical Society Series B, 57, 1, 289-300, 1995.

[5] Sean D. et al. "GEOquery: a bridge between the Gene Expression Omnibus (GEO) and BioConductor"; Bioinformatics, 23, 1, 1846-1847, Jul 2007.

[6] Smyth, Gordon K. "Limma: linear models for microarray data"; In Bioinformatics and computational biology solutions using R and Bioconductor (Springer New York), 397-420, 2005.

[7] R documentation. "Adjust P-values for Multiple Comparisons".

[8] Thomas P et al. "PANTHER: a library of protein families and subfamilies indexed by function"; Genome Research, 13, 9, 2129-2141, 2003.

[9] Kelley et al. "The Phyre2 web portal for protein modeling, prediction and analysis"; *Nature Protocols* 10, 6, 845-858, 2015.

[10] J Yang, et al. "The I-TASSER Suite: Protein structure and function prediction"; Nature Methods, 12, 1, 7-8, Jan 2015.

[11] Biasini M et al. "SWISS-MODEL: modelling protein tertiary and quaternary structure using evolutionary information"; Nucleic Acids Research, 42(Web Server issue), W252-W258, July 2014.

[12] McGuffin LJ et al. "IntFOLD: an integrated server for modelling protein structures and functions from amino acid sequences"; Nucleic Acids Research, 43(Web Server issue), W169-W173, Jul 2015

[13] Szklarczyk D et al. "STRING v10: protein-protein interaction networks, integrated over the tree of life"; Nucleic Acids Research, 43(Database issue), D447-D452, Oct 2014.

[14] Hochberg Y. "A sharper Bonferroni procedure for multiple tests of significance"; Biometrika, 75, 4, 800-803, July 1988.

[15] R Documentation. "Table of Top Genes from Linear Model Fit".

[16] Altschul SF et al. "Basic local alignment search tool"; J Mol Biol, 215, 3, 403-10, 1990.

[17] Shen Y et al. "EOL-1, the Homolog of the Mammalian Dom3Z, Regulates Olfactory Learning in C. elegans"; The Journal of Neuroscience, 34, 40, 13364-13370, Oct 2014.

SESSION

NOVEL STUDIES

Chair(s)

TBA

Recurrent Breast Cancer Treatment via Rapid Pain-Guided Experimentation

Jane Eyre[1]**, Steve Richfield IEEE # 41344714**[2]
[1] Owner of NormalBodyTemperature.co.uk TudorJane@gmail.com
[2] Owner of FixLowBodyTemp.com Steve.Richfield@gmail.com

Abstract – *Recurrent breast cancers present a special challenge. Spawning from mutated surviving cancer cells or oncogenic cysts, they tend to be aggressive and malignant, and many have mutated to circumvent measures that killed an earlier tumor, so survival rates are negligible.*

There is a brief window of time, starting when tumors first become painful, and ending when they have grown enough to do sufficient damage to surrounding tissues so that pain cannot be quickly stopped. When tumors are exquisitely sensitive to their internal pressure, it is possible to quickly test prospective adjuvant therapies and receive immediate feedback as to their efficacy in the form of prompt reduction or elimination of pain.

This paper concentrates on rapidly adjusting body temperature, in the hope and expectation that a temperature can be found where a dormant component of the immune system will awaken and attack the cancer.

Keywords: stage 4, recurrent breast cancer, body temperature, thermotherapy, pain, low-dose challenges

1 Introduction

Jane, a breast cancer "survivor" residing in the UK, spends much of her time working with people who are trying to survive their own cancers. Jane had fixed all of the factors known to contribute to a recurrence, including cleaning up her diet, taking all cancer-fighting supplements purported to help, raising her body temperature, etc. So, when a tumor and several enlarged lymph nodes developed nearly seven years after she thought she had been "cured", she knew she was in a fight for her life, because there was literally nothing else left to fix. Further, just two days after their discovery, Jane's tumor became severely painful.

Desperate times beget desperate measures. Recurrent cancers often quickly kill their hosts - before there is even time to receive competent cancer treatment. The majority of the work described herein was done before Jane's first available oncology appointment.

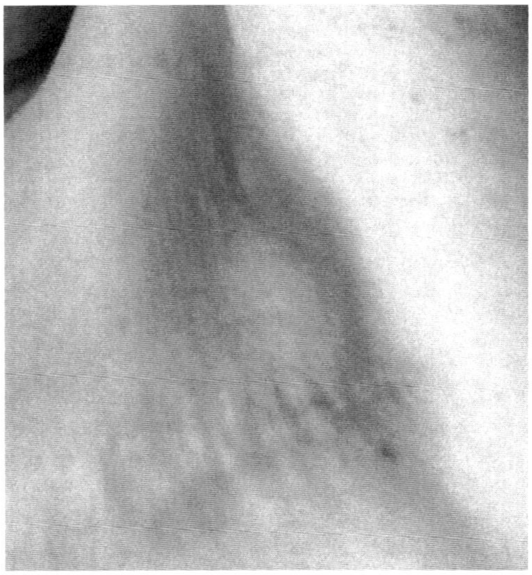

Fig. 1: Metastatic tumor in left armpit shown in actual size. In 2 weeks this tumor shrunk leaving no visible bump. Previous mastectomy scar is shown in lower left.

Steve Richfield is currently the only known Central Metabolic Control Systems (CMCS) therapist, specializing in changing various CMCS set-points, most commonly body temperature. This project is unusual because it involves finding previously unknown set-points, and altering Jane's daytime temperature to values other than the usual 37°C=98.6°F.

2 Recurrent Breast Cancer Biology

The most commonly accepted theory is that surviving cells mutate to form tumors that are more problematical than the original tumor. Here, tumors are often found in clusters of successively mutated cells that sometimes respond differently to treatment.

A very common treatment for breast cancer is a mastectomy (removal of the breast along with many lymph nodes). Without lymph nodes, exudate from surrounding tissue collects in a cyst the body forms to hold them. Then, one of two things happens:

1. the body is NOT able to extract small molecules from the exudates fast enough to keep up, so the cyst grows without limit, becoming lymphedema, or

2. the body IS able to extract small molecules from the exudates fast enough to keep up, so the exudates becomes concentrated in a sort of primordial soup in an oncogenic cyst, from which new forms of life emerge – including aggressive cancers.

The sad irony here is that the patients who do NOT contract lymphedema, often because they are very careful not to use their chest muscles on that side, are seen as the "lucky" ones, when it is they who proceed without symptoms until they develop an almost universally fatal tumor. The patients who develop lymphedema receive treatment to get their cysts drained and flushed, and so they have better survival prospects.

Jane had no lymphedema before her tumor developed, so it came as no surprise to learn her worst tumor was covering the top of her oncogenic cyst. However, upon ultrasonic examination it was found that Jane had a cluster of three tumors. Hence, the genesis of this particular cancer won't be determined until differences in response to treatment are seen, or not seen.

3 Immune Subsystem Selection Biology

Immune systems operate via complex enzymes - the MOST complex enzymes that it is possible to construct given 100 millions years of evolution. The thing that establishes the upper limit to the complexity of chemical reactions is temperature range – as reactions become more complex, their temperature range becomes narrower, with the result that some parts of our immune systems only operate at one specific temperature with no remaining "range". Hence, components can be individually selected by adjusting body temperature.

As a result, people must have changing temperatures for everything to work right, with most healthy people sleeping at 36.3°C≈97.4°F and rising to 37°C=98.6°F during the day.

Some (rare) people with cancer spontaneously run fevers which usually kill their tumors. However, it is suspected that these fever temperatures may frequently change to activate the multiple mechanisms needed to kill the most robust tumors and invaders.

There are good reasons for saving some immune subsystems for when all else has failed - to avoid "helping" tumors and invaders to evolve to the point of being able to avoid an immunological attack. To avoid such evolution, our CMCS does NOT ordinarily activate some last-resort parts of our immune systems. This is why our CMCS waits for a major attack of some sort before resorting to running a fever, instead of cycling to fever temperatures all of the time.

4 How Temperature Set-points Work

Everyone has an assortment of the same temperature set-points distributed approximately every 0.33°C=0.6°F from each other, somewhat like a rotary switch with stable detents (catches), similar to an old fashioned television tuner channel selection switches where each channel is different. Each set-point temperature works differently. Most healthy people end up using every other set-point, e.g. 36.3°C≈97.4°F and 37°C=98.6°F, while skipping the intervening set-point at 36.6°C≈98.0°F, much like alternating television channels are assigned. People's brains decide which set-point to use at any moment in time, switching abruptly between them as needed. With a little practice, most people can sense these changes as slight momentary chills or flushes that have become so familiar that they are no longer noticed. Absent extreme circumstances, most people spend nearly all of their time at just two or three of their set-points.

Set-points are easily measured by adjusting the ambient temperature for perfect comfort and measuring body temperature. That many people have found the same set-points confirms that these are the same for everyone.

With some practice, most people can learn to observe physiological clues and guesstimate their temperatures to within ~±0.1°C≈0.2°F without using a thermometer.

Various things commonly go wrong with temperature regulation, the most common being central hypothermia. Therein, superstitious learning, most often from previous general anesthesia having "frightened" the CMCS into never again operating at the same temperature, makes it impossible to utilize much of the normal range of body temperatures. This "red tagging" blocks the use of a particular set-point, and usually all other set-points above it, thereby disabling many of the capabilities of a person's immune system. This was clearly the problem for Jane when she developed her original breast cancer seven years earlier.

The present project involved searching for previously unknown set-points, which turned out to be at 37.4°C≈99.3°F and 37.7°C≈100°F.

Note that the Fahrenheit temperature scale was originally created so that 100°F would be normal body temperature. Perhaps the German physicist Daniel Gabriel

Fahrenheit (1686–1736) lived much of his life at the 37.7°C≈100°F set-point, which might help explain why he only lived to be 50.

These two previously unknown set-points were found by Jane when she raised her temperature in a sauna up to 38°C≈100.5°F, and then ever so gradually lowered her temperature over the course of several hours, as she carefully watched for indications of an active set-point. The primary indication turned out to be feelings of thermal comfort.

5 History – Round 1

From 2005 to 2009 Jane had many symptoms of low thyroid function and increasingly had cachexia (wasting syndrome). Blood tests revealed nothing wrong. June 2009 Jane was diagnosed with a 4cm ductal invasive carcinoma of the breast. Jane still had cachexia, which her oncologist said was unrelated. Jane's treatment was a mastectomy and 6 bedridden rounds of chemotherapy. These treatments were traumatizing to Jane. Fearing for her life, Jane decided not to have radiotherapy or aromatase inhibitors. Jane still had cachexia. Blood work still revealed nothing and Jane was becoming dangerously ill. Suspecting that her abnormally low body temperature might be the underlying cause for all of her problems, Jane contacted author Steve Richfield for assistance in normalizing her body temperature. Jane's cachexia and other symptoms soon resolved.

6 History – Round 2

Nearly seven years later, Jane began feeling discomfort in her armpit in the region of her previous mastectomy. Two months passed before serious pain began. Initially the pain was barely perceptible, a little twinge, a little tightness, then the discomfort prompted some gentle manipulation which revealed two lumps. One lump was located under Jane's left armpit and measured about 3cm wide and 1.5cm deep. The other was a lump which could float across a rib. This lump was 1.5cm wide and 0.5cm deep. Other "minor aches and pains" could possibly have been other metastases. The pain was becoming unbearable, prompting Jane to seek help from the medical system and author Steve Richfield.

Jane had not stuck with her original temperature maintenance program. She had allowed her daytime temperature to drop as part of a plan to save money by not adequately heating her home. Living life between set-points can be even more dangerous than living life at the wrong set-points, because at a wrong set-point, at least some part of your immune system has been selected to be active. Now Jane was probably suffering from the expected consequences.

A careful review of Jane's past lab tests disclosed a thyroid panel showing that her TSH, FT4, and T3 were all nearly at the bottom of their respective ranges. Since Jane was clearly metabolically challenged (which is one of the suspected causes for her present cancer), the only apparent explanation seemed to be some sort of pituitary or hypothalamic malfunction for which there is no known direct intervention. Jane discovered she felt better when she tried some low-dose T4, but full T4 supplementation is well known to short-circuit thermotherapy efforts by forcing temperatures to 36.6°C≈98.0°F.

7 The Plan

The plan started out simply – to stay alive and relieve pain long enough to secure competent help, possibly including surgery to remove the painful tumors. NSAID pain relievers had no apparent effect. Jane immediately went on a "zero carbohydrate" ketogenic diet, but its effects are too slow to evaluate. Jane started eating apricot kernels on hand, only to develop symptoms of cyanide poisoning, so she reduced her dosage. To help generate the heat needed to maintain a higher temperature, Jane utilized T2, T3, and T4 supplements. Jane ordered some dichloroacetate (DCA), but it apparently got hung up in British customs for several days. Taking the DCA seemed to reduce the pain, but there were too many other things happening to be sure.

Jane entered her sauna and observed how she felt at various body temperatures. It became apparent that at some specific temperatures above 37°C=98.6°F the pain was greatly reduced and often completely gone.

With temperature manipulation being the only thing that seemed to help, efforts focused on temperature manipulation.

8 Challenges

No one knows what parts of the immune system are activated by various temperatures. This probably changes from one person to the next because people have differing immune systems. Further, killing a difficult tumor could require cycling between several unknown temperatures.

Supporting this theory, raising Jane's body temperature to 37.4°C≈99.5°F was first observed to eliminate pain – but it gradually returned until switching to 37.0°C=98.6°F or 37.7°C≈100°F. Fighting cancer is clearly a complex time-dependent process.

To address this complexity, a pragmatic approach of "if it hurts, change the body temperature" was adopted. After just one day, several temperature set-points seemed to develop their own individual personalities. However, both people and tumors are unique, so take Table 1 as an example and NOT as a reference.

Table 1: Jane's Stable Oral Temperature Set-points vs. Observed Pain

35.9°C≈96.7°F	PAIN
36.3°C≈97.4°F	Sleep without pain.
36.6°C≈ 98.0°F	Increases pain.
37.0°C=98.6°F	Doesn't seem to change pain.
37.1°C≈98.8°F	Gradually decreases pain.
37.4°C≈99.5°F	Usually decreases/eliminates pain.
37.7°C≈100°F	Always decreases/eliminates pain.

Yoshimizu[1] discusses (on page 82) Dr. Frank T. Kobayashi getting improvement in 70% of his cancer patients by briefly heating them to 39-40°C≈102-104°F for two hours while simultaneously administering just 5-10% of the usual doses of chemotherapy. This was presumably done without real-time experimentation with these or other temperatures.

9 Misleading Effects

There are several ways to briefly reduce cancer pain and potentially mislead rapid experimentation, which should be kept in mind to avoid being misled.

- Lower metabolism, e.g. from heating, restricted diet, etc., can retard tumor growth and stabilize tumor pressure.
- Dehydration can cause water to be removed, resulting in reduced tumor pressure.
- Lower blood pressure, e.g. from blood pressure medications, can reduce how much the heart pumps up tumors.

Hence it is important to also watch other indicators like tumor softness, sensitivity to applied pressure, tumor size, etc.

10 An Epiphany

Maintaining desired temperature proved to be increasingly more difficult because Jane developed extreme hot flushes that drenched her. Suspecting her CMCS was alerting over some unknown bad situation, a survey of temperatures around her body showed that her body was 0.65°C≈1.2°F cooler than her head.

This means Jane's temperature measurements were of Jane's head and NOT her body where her tumors are situated. Adjusting for this temperature difference, it became clear that Jane's body, but not her head, has been in the 35.something°C≈95.something°F range which is typical of the vast majority of cancer cases.

Jane's central hypothyroidism that would be expected to impair heating of her organs, as evidenced by her low FT4 and T3, Jane tried heating her body while simultaneously cooling her head. This worked amazingly well, with Jane's head then easily rising to 37.1°C≈98.8°F without medication, the difference between head and body temperature cut in half, and hot flushes becoming barely noticeable.

Understanding and effectively addressing Jane's central hypothyroidism will be needed before Jane can stay alive without dressing for a Siberian winter while fanning her face. Conventional treatment via T4 supplementation doesn't work because it causes her temperature to drop. The following tests are now planned:

- TRH Challenge to assess pituitary function
- IGF-1 to assess pituitary function
- MRI to look for pituitary and hypothalamic tumors

However, it is likely that no laboratory test will find a malfunction, because the malfunction may be just another superstitious learning artifact from the same exogenous cause (e.g. general anesthesia as a child) that caused her central hypothermia that lead to her original breast cancer.

Kokolus, et al, have observed that chemotherapy often causes subsequent hot flushes. Perhaps some chemotherapy agent(s) cause damage to the hypothalamus or pituitary leading to central hypothyroidism. This would not be surprising because chemotherapy is well known to cause cognitive impairment, sometimes called "brain fog", and there are many delicate sensing neurons in the hypothalamus and pituitary. Someone should do the research to identify the chemotherapy agents used on patients who subsequently developed hot flushes, as this appears to be a dangerous but previously unrecognized side effect of some common forms of chemotherapy.

11 Turnaround

Starting her day with a brief session in a cabinet that only heats her body while a fan cools her head, and upping her T2 intake, Jane now feels GREAT with her only remaining symptoms being her now-painless shrinking tumor and some mild liver discomfort, hopefully from processing lots of recently killed cancer cells. Jane's head and body are now at the same 37.1°C≈98.8°F temperature, even after she left her cabinet. Jane now has the energy to clean her house and catch up on her many chores that have gone undone while she has been so sick.

12 Examination

Just before the cutoff time for publication, Jane finally managed to have an oncology examination including an ultrasound scan of her tumor. Jane's "tumor" is actually a cluster of three tumors. Further, one of the tumors extends between her left-side ribs and into her body, and hence is inoperable. Irradiating her heart would not be good, and besides, there are probably other tumors in remote locations.

Chemo would be expected to have poor results against such an aggressive tumor in a patient who has already received chemo for a past tumor, and who has reacted so badly to chemo in the past. Jane's present thermotherapy is clearly working better than any of the other conventional cancer treatment could work.

13 Unexpected Encouragement

The day after her oncology exam, Jane discovered an apparent small metastatic tumor in the cuticle of her thumbnail. Having already lived longer than she expected when her adventure started, a new tumor HERE, on the coldest part of her body rather than at a much more problematical location, further supports the proposition that temperature is holding other tumors at bay and/or is killing them.

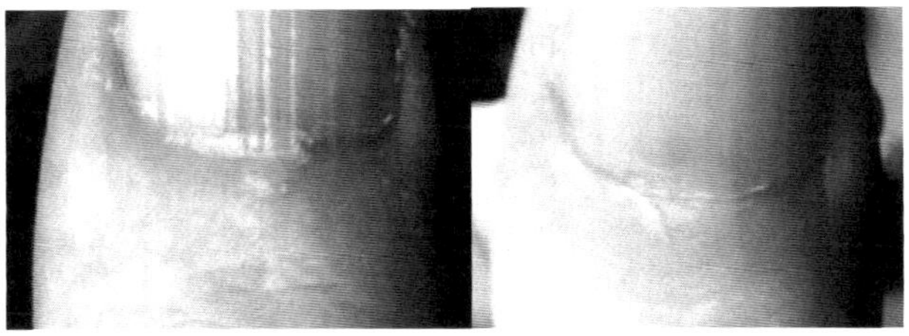

Fig 2: Probable early metastatic tumor in thumbnail cuticle, when noticed and two days later, shown twice actual size.

The wonderful thing about this particular tumor is that it greatly facilitates further experimentation, e.g. testing at difficult-to-attain temperatures, before considering prospective techniques to achieve those temperatures across an entire body.

Yoshimizu[1] reports (on page 71) that most cancer cells can survive immediate destruction to ~42°C≈108°F whereas healthy cells survive to 47°C≈117°F, providing a thermal window to destroy cancer cells. However, the temperature required to kill cancer cells is doubtless dependent on the specific mutations involved. Note that this report addresses cellular survival and not functioning, as these temperatures exceed those known to cause brain and other damage. Hence, whole body incineration of tumors without relying on immune systems does not appear to be practical.

14 Results

The primary reason for rapid real-time experimentation is to **quickly** learn what works, without introducing the traditional "noise" of individual patient survival statistics by counting noses and headstones. This means that the usual measure of results, patient survival statistics, is rapidly becoming an obsolete measure. However, patient survival remains VERY important for those involved, most especially the individual patients.

Jane's tumors have clearly shrunk – to around half their peak size, or 1/8th their original volume as of this writing. However, shrinkage has not been steady. When she wakes up in pain, she observes her tumors have increased in size, but they then shrink again when she raises her temperature and eliminates the pain. Measurement is made by placing a piece of tape over a tumor and pushing a pen into the tape as she goes around the tumor.

More will be learned by the time this paper is presented at WORLDCOMP. With luck, Jane will be able to attend the WORLDCOMP conference and discuss the events presented here from her first person point of view.

15 Conclusions

15.1 Viability of Rapid Experimentation

Rapid experimentation clearly works for treatments like thermotherapy and some chemotherapy, where early results can be immediate.

Rapid experimentation probably works best on aggressive tumors that quickly "recover" from successful treatments.

Rapid experimentation clearly does NOT work for treatments that have slow response times, like dietary interventions.

Rapid experimentation testing of new substances on several patients might facilitate a screening process for prospective new therapies.

15.2 Viability of Thermotherapy

Short-term thermotherapy without precise temperature control has been studied as a cancer treatment by Yoshimizu[1]. At risk of oversimplification, he seems to say that its greatest value is as an adjunct to chemotherapy, where good results can be secured with only 10% of the usual doses of chemotherapy. Most oncologists seem to think that surgery is usually needed, as otherwise cancer cells can hide within tumor structures.

Thermotherapy, as part of a comprehensive plan to correct erroneous body temperature, has been successfully practiced by many people.

Thermotherapy, as an emergency treatment for the progression and extreme pain often associated with cancer, seems to be proven by our experiences reported herein.

15.3 Watch for Central Hypothyroidism

Central hypothyroidism is a somewhat rare condition where the brain fails to send hormonal signals for the thyroid to sufficiently activate. Central hypothyroidism is evidenced by simultaneously-low TSH and FT4, that may both be above the lower limits of their respective reference ranges, whereas other hypothyroidism is evidenced by elevated TSH, often accompanied by low FT4. As a result, TSH screenings for hypothyroidism completely miss cases of central hypothyroidism. However, breast cancer is also a somewhat rare condition, and the two might be related, as in Jane's situation. Until more is known, it seems prudent to perform a thyroid panel test on patients where recurrent cancer is suspected, and look for TSH and FT4 **both** being low, even if they are both within their respective reference ranges.

15.4 T2 Research Needed

There are lots of anecdotal postings on the Internet regarding the efficacy of the thyroid hormone T2. T2 is clearly not inert, as many endocrinology textbooks would have you believe. Complicating this situation is that T2 comes in several forms, as there are 4 sites on which to attach 2 iodine atoms, and competing suppliers advertise the benefits of their differing forms of T2. Research is needed to better understand the action of various forms of T2, to better guide its therapeutic use.

15.5 New Type of Chemotherapy Damage

Large numbers of chemotherapy patients developing hot flushes, coupled with the understanding of Jane's hot flushes as explained here, strongly suggest that some chemotherapy agent(s) may be causing a specific sort of brain damage that causes body temperature disturbances and/or central hypothyroidism - that might lead to future cancers. Someone should do the research to identify and ban such substance(s), though in Jane's case TAC has already been banned in much of the world – but not in the UK.

For Jane's initial breast cancer, she had the chemotherapy cocktail TAC, an acronym for:

- **T** - docetaxel (also called **T**axotere®)
- **A** - doxorubicin (also called **A**driamycin®)
- **C** - **C**yclophosphamide.

15.6 Painful Lumps as an Emergency

Emergency room personnel are ready to help people who develop chest pains, but they are NOT (yet) ready to help future cancer patients who have just developed their first painful lump(s). If hospitals were to prepare for this by having a plan to start evaluating prospective adjuvant therapies, they could utilize the brief window of time when tumors are exquisitely sensitive to pressure to determine what treatments will best kill the cancer. Immediate experimentation, even in the absence of high-tech testing, can greatly alter the future course of tumors.

15.7 Questionable Ethics of Removing Large Numbers of Lymph Nodes

Removing large numbers of lymph nodes, without transferring replacement lymph nodes (from the groin), creates a deadly oncogenic cyst and/or causes lymphedema. Transferring lymph nodes requires a special skill that is not commonly available. As a result, there are presently far more mastectomies with lymph node removal being performed, than there are doctors to transfer replacement lymph nodes, so only patients who have already progressed to the edge of death qualify for these procedures. Since the patients who do NOT develop lymphedema often instead develop quickly-fatal tumors and promptly die, this sort of "triage" greatly reduces the number of patients receiving this procedure - by killing the stronger patients.

The methods presented herein provide a new approach to sparing lymph nodes, despite the likelihood of them harboring cancer cells, because patients can now KNOW that they have a good adjuvant therapy, proven to work on their own tumors and surviving cancer cells, to use immediately and/or when their tumors have been surgically removed. With this, a surgeon can abandon procedures intended to remove all cancer cells, and instead only remove tumors, relying on the already-proven adjuvant therapy to reliably kill all remaining cancer cells once surgery has been completed. As a result, only those lymph nodes harboring tumors need be identified and removed.

16 Warnings

Alternative health literature is full of outrageous claims of new cancer "cures". However, there are nearly as many forms of cancer as there are people who have cancer, so there probably never will be a cure-all for cancer. There are LOTS of things that seem to help – for a while until the cancer mutates to circumvent ongoing measures. If you wonder whether something you read about really worked, your best first step is to contact the person who made the claims, and see if that person is still alive.

There are many people who believe the "gold standard" for curing cancer is restoring normal operation of the immune system – which Jane did with her initial cancer. Jane's experiences challenge this simplistic belief, because while her body temperature was usually a little low, it was also often 37°C=98.6°F – especially whenever she drove herself around with the heater in her car turned up high. Perhaps if UK's winter had not been quite so cold this year, Jane's immune system might have kept her cancer at bay for another month, and then killed off her cancer (yet again?) when summer heat arrived. Jane will surely pay more attention to these "little" details in the future.

You will find no cure-alls here, but you will find techniques that will facilitate each person's rapid search for their own individual cures. In the process, these methods can quickly alleviate the excruciating pain associated with cancer, though at the cost of some discomfort from raising body temperature, possibly higher than it has ever before been.

17 Future of this Methodology

For the first time, this methodology will facilitate the side-by-side comparison of competing cancer treatments ON THE SAME TUMORS. This will facilitate the search for genetic markers preferring one treatment over another, instead of searching for markers for each isolated treatment without any ability to compare competing approaches.

Using genetics to determine in advance which adjuvant therapy will work best, instead of only determining which adjuvant therapy will work at all, should propel the practice of cancer treatment beyond present-day limitations.

It is hoped that the methods presented herein will bring an end to the practice of using adjuvant therapies that haven't been pre-tested on each patient, which often does considerable harm without killing tumors.

18 References

[1] Nobuhiro Yoshimizu. "***The Fourth Treatment for Medical Refugees***". RichWay International, Inc, 2009. http://www.bio-mats.com/the-fourth-treatment-for-medical-refugees/table-of-contents

This hyperlink provides access to the entire book in 7 languages. Therein, case studies of 17 patients treated with thermotherapy show some whose temperatures spontaneously rose much as Jane has observed, despite the lack of closed-loop temperature control as Jane used.

[2] Canadian Cancer Society "***Types of chemotherapy***" http://www.cancer.ca/en/cancer-information/diagnosis-and-treatment/chemotherapy-and-other-drug-therapies/chemotherapy/types-of-chemotherapy/?region=on

Should you be considering chemotherapy, this overview of chemotherapy agents will facilitate figuring out which are most likely to be of benefit in your particular situation.

[3] Kathleen Kokolus, Chi-chen Hong, and Elizabeth Repasky. "***Feeling too hot or cold after breast cancer: Is it just a nuisance or a potentially important prognostic factor?***". Int. J. Hyperthermia. 2010. http://www.ncbi.nlm.nih.gov/pmc/articles/PMC301237

This paper discusses how hot flushes from prior chemotherapy present a major challenge for breast cancer "survivors". Our experience shows that a substantial temperature difference between body and head can cause hot flushes, which are reduced or eliminated by dressing very warmly while exposing the head.

[4] Karl Groth, Theodore Kelly, Todd Westerbeck, and Gary Blick. "***Treatment of human herpes viruses using hyperthermia***". U.S. Patent 6,415,797, 2002. http://www.google.co.uk/patents/US6415797

Pay particular attention to the "EXAMPLES" section in this U.S. thermotherapy patent.

SESSION

POSTER PAPERS

Chair(s)

TBA

A Gain Compensation Algorithm for Hearing Aid using the Voice Activity Detection

Sang-Kyun Kim[1*], Sang-Ick Kang[1], Young-Jin Park[2], Jang-Woo Kwon[3] and Sangmin Lee[1]

[1]Department of Electronic Engineering, Inha University, Incheon, 402−751, South Korea
[2]Korea Electrotechnology Research Institute (KERI), 111 Hanggaul ro, Sangrok Gu, An-San shi, Kyunggi Do, 426-170, South Korea
[3]Department of Computer Engineering and Information, Inha University, Incheon, 402−751, South Korea
[*]Contact Author
greenwhity@nate.com[*], rkdtkddlr@gmail.com, yjpark@keri.re.kr, jwkwon@inha.ac.kr, sanglee@inha.ac.kr

The type of the submission: Extended Abstract/Poster Paper

Abstract - *In this paper, we propose a novel approach to improve the speech quality of the hearing aid in noisy environment. The conventional gain control algorithm suppresses the noise of input signal, and then the part of wide dynamic range compression (WDRC) amplifies the undesired signal. The proposed algorithm controls the gain of hearing aids according to speech present probability by using the output of a voice activity detection (VAD). The performance of the proposed scheme will be evaluated under various noise conditions by using objective measurement*

Keywords: Gain Compensation, Voice Activity Detection, Noise Suppression, Wide Dynamic Range Compression

1 Introduction

The hearing aid is an electroacoustic device that amplifies the input sound. The input signal from microphone is processed by several algorithms. The noise suppression (NS) algorithm has an impact on the sound quality of the hearing aid, directly [1]. Another important technique at the hearing aid is the wide dynamic range compression WDRC algorithm that nonlinearly amplifies the input signal according to level of input signal. The character of hearing impairment can't hear the low level sound, but the enough loud sound is possible to hear. For this reason, the input signal is controlled by WDRC according to the level of input signal inverse proportionally. Unfortunately, the NS algorithm performs more early than the WDRC algorithm, so the suppressed noise signal is increased [2].

In this paper, we propose a novel approach for the adaptive gain compensation by employing the voice activity detection (VAD). First of all, the input signal is distinguished by the VAD scheme which determines the speech presence. The noise component in the input signal is suppressed by NS algorithm, and then the enhanced signal selectively is processed by the WDRC technique according to the result of the VAD.

2 Review of Conventional Algorithm

Among the techniques for hearing aid, the algorithms of high correlation with gain are the NS and WDRC algorithm. Fig. 1 represents a typical flowchart of the hearing aid algorithm.

2.1 Noise Suppression Algorithm

In the time domain, it is assumed that the environmental noise signal $n(t)$ is added to the clean speech signal $s(t)$, with their sum being denoted by $y(t)$, which is called the noisy speech signal. They are transformed by a short-term fast Fourier transform (FFT) as follows:

$$Y(k,l) = S(k,l) + N(k,l) \tag{1}$$

where Y, S and N are FFT coefficients of the noisy speech, clean speech and noise respectively. The frequency bin and frame index are expressed as k and l respectively. The *a posteriori* signal-to-noise ratio (SNR) $\gamma(k,l)$ and the *a priori* SNR $\xi(k,l)$ are given by [3]

$$\gamma(k,l) = |Y(k,l)|^2 / \lambda_N(k,l) \tag{2}$$

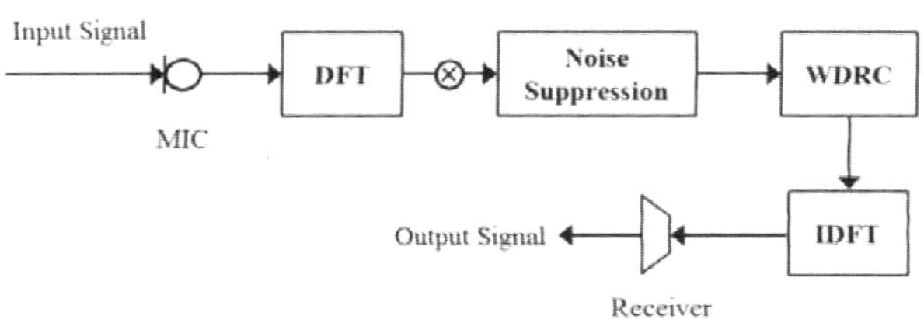

Fig. 1. Block diagram of hearing aid

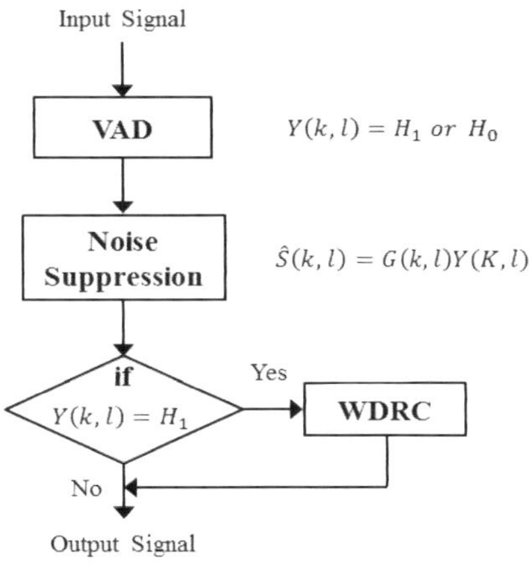

Fig. 2. Flowchart of proposed algorithm using the VAD for hearing aid

$$\xi(k,l) = \lambda_S(k,l) / \lambda_N(k,l) \quad (3)$$

where $\lambda_S(k,l)$ and $\lambda_N(k,l)$ denote the speech and noise variances, respectively. The speech spectral is amplitude estimate at the minimum mean-square error (MMSE) estimator that is defined by [4]

$$\hat{S}(k,l) = G(\xi(k,l), \gamma(k,l))Y(k,l) \quad (4)$$

where $G(\xi(k,l), \gamma(k,l))$ is gain function.

2.2 Wide Dynamic Range Compression

Through the WDRC algorithm, the gain of the estimated speech signal is assessed by power level as follows.

$$O(k,l) = \begin{cases} w_1 \cdot \hat{S}(k,l) & \text{if} \quad \hat{S}(k,l) < T_1 \\ w_2 \cdot \hat{S}(k,l) & \text{if} \quad T_1 \le \hat{S}(k,l) < T_2 \\ \hat{S}(k,l) & \text{otherwise} \end{cases} \quad (5)$$

where a weight w and a threshold T are dependent on the hearing ability and range, respectively.

3 Proposed Algorithm

The proposed algorithm discriminates between the speech presence and absence frame before the NS algorithm by employing VAD. Fig. 2 presents the block diagram of proposed algorithm. In this paper, the used VAD is a statistical model-based method. Given two hypotheses, $H_0(i)$ and $H_1(i)$, that respectively indicate speech absence and presence in the noisy spectral component, it is assumed that

$$H_0 : Y(k,l) = N(k,l) \quad (6)$$

$$H_1 : Y(k,l) = S(k,l) + N(k,l) \quad (7)$$

The likelihood ratio $\Lambda(k,l)$ computed in each frequency bin as follows [3]:

$$\Lambda(k,l) = \frac{1}{1+\xi(k,l)} \exp\left\{ \frac{\gamma(k,l)\xi(k,l)}{1+\xi(k,l)} \right\} \quad (8)$$

The decision rule of the voice activity is given by

$$\Lambda(l) = \prod_{k=1}^{M} \Lambda(k,l) \underset{H_0}{\overset{H_1}{\gtrless}} \eta \quad (9)$$

4 Conclusions

In this paper, we have proposed a novel approach to improve the speech quality of the hearing aid in various noise environments. Through the VAD technique, we presuppose better performance than the conventional method. Through the VAD technique, we presuppose that the results of the experiment will be positive.

5 Acknowledgments

This research was supported by the KERI Primary Research Program through the Korea Research Council for Industrial Science & Technology funded by the Ministry of Science, ICT and Future Planning (No. 16-12-N0101-44) and was supported by Basic Science Research Program through the National Research Foundation of Korea(NRF) funded by the Ministry of Education(2010-0020163).

6 References

[1] B. Edwards, “The future of hearing aid technology,” *Trends in Amplification*, vol. 11, no. 1, pp. 31-46, Mar. 2007.

[2] C. M. Lee, S. H. Bae, J. H. Kim, and N. S. Kim, “Spectro-temporal filtering based on soft decision for stereophonic acoustic echo suppression,” *J. Commun. Networks (JCN)*, vol. 39C, no. 12, pp. 1346–1351, Nov. 2014.

[3] S.-K. Kim and J.-H. Chang, "Voice activity detection based on conditional MAP criterion incorporating the spectral gradient," *Signal Processing*, vol. 92, no. 7, pp. 1699-1705, Jul. 2012.

[4] Y. Ephraim and D. Malah, “Speech enhancement using a minimum mean-square error short-time spectral amplitude estimator,” *IEEE Trans. Acoustics, Speech and Signal Processing*, vol. ASSP-32, no. 6, pp. 1190-1121, Dec. 1984.

Broadening the Scope of Computer Science Education: Introducing and Incorporating Biodiversity Informatics education to a Computer Science Curriculum

Nazli W. Hardy [1], **Christopher R. Hardy** [2, 3]
[1]Computer Science, Millersville University, Millersville, PA, USA
[2]Biology, Millersville University, Millersville, PA, USA
[2]James C. Parks Herbarium, Millersville University, Millersville, PA, USA

Abstract - *As the field of computer science continues to be pervasive in in every other field of study, it is necessary to introduce and incorporate multidisciplinary aspects of those fields to the curriculum. At Millersville University, the "freshman year experience course" for computer science students gives incoming students an overview of the wide spectrum of the study of computer science (networking, human-computer interaction, gaming, bioinformatics, etc.). This introduction can be developed further by giving interested students the opportunity to engage in-depth in interdisciplinary, independent studies with professors who are experts in the related fields. Biodiversity informatics is a practical and important branch of computer science that can be described as the collection, curation, analysis, and interpretation of information regarding biodiversity – with the aid of computing power. In addition, biodiversity informatics plays a critical role in the conversation of species in light of climate change. Thus biodiversity informatics has an importance in expanding the role of computer science for undergraduate and graduate students. Since 2007, faculty from the departments of Computer Science and Biology at Millersville University have worked with eleven computer science undergraduate students, in a research capacity, to develop NatureAtlas, NatureAtlas is an application that has eight primary portals: plants, birds, fishes, fungi, herps, invertebrates, mammals, zooplankton. The plant portal is the most well populated because of to a) the sheer number of species, b) students at six universities have used the plant portal in Plant Systematics classes, and c) nature societies and biological preserves have used it for citizen scientist-led conservation projects (bioblitzes). The authors welcome continued and multidisciplinary collaborations within other education and conservation institutions in this endeavor.*

Keywords: *Multidisciplinary Studies, Biodiversity Informatics, Computer Science Education*

Predicting the Function of Hypothetical Protein PANDA_003700 using Computational Analysis Methods

Cameron Bixby and Padmanabhan Mahadevan*

Dept. of Biology, University of Tampa, Tampa, FL 33606
*To whom correspondence should be addressed (pmahadevan@ut.edu)

Abstract

The majority of the gene products produced after an organism is sequenced are proteins whose function is not known, called hypothetical proteins (HPs). Proteins that are predicted from nucleic acid sequences only and proteins with unknown functions are considered hypothetical proteins (Lubec et al., 2005). Therefore as large amounts of hypothetical proteins are discovered from genomic sequencing, they will continue to enter the spotlight of many studies in the Bioinformatics and Genomics field. About half of the proteins in most genomes are candidates for HPs (Lubec et al., 2005). Therefore, determining the function of the HPs is very important when trying to complete the genomic and proteomic information of a sequenced organism. HPs are observed across a variety of phylogenetic lineages but their functions are not characterized (Galperin & Koonin, 2004).Therefore, the challenge to characterize the function of HPs using experimental and computational methods has become more important in genomic studies.

Typically, the work dedicated to discovering the functions of HPs can be separated into two parts: prediction of protein function through its sequence and prediction of the 3-D structure of an HP. In terms of predicting the function of a HP through its sequence, researchers will use computational methods in order to compare their HP against functional proteins in hope of high sequence identities. In finding the similarities between sequences, researchers can infer the function of the protein, explore protein families, and evolutionary relationships (Lubec et al., 2005). The most common tool in calculating sequence similarity is the Basic Local Alignment Search Tool (BLAST) which has a version that can blast a protein query against a database of proteins. Exploration of various protein families to see if the HP shares any common evolutionary origin is another route taken by researchers in order to gather more information on their HP. Protein families are sets of protein regions which share a significant degree of sequence similarity (Punta et al, 2011). Therefore using databases like Pfam can display various relationships between a HP and other functional proteins. It is also important to mention that protein domains are also considered another area in which a researcher can use the HP sequence to discover its domains. Protein domains are viewed as the basic components of proteins and from this it helps determine the functional characterization (Veretnik et al., 2004).

The sequence of a hypothetical protein can provide a lot of insight in terms of the prediction of the protein structure itself which can then further help determine the function of the HP. This ties in with the goal of structural genomics which is to create a complete inventory of protein folds/structures that can help predict functions for all proteins (Mittl & Gütter, 2001). One way to determine the 3-D structure of a protein is by either x-ray crystallography or NMR, then the structure can be compared against other structures in a protein database (Zarembinski et al., 1998). However, those experimental methods are usually difficult, complex, and time consuming. Therefore, with limited experimental models of proteins to compare with a HP, homology modeling has become a reliable way to determine the 3-D structure by using a HP's amino acid sequence. It is important to mention that homology searches are more accurate if the sequence similarity between the HP and the homolog of another known protein is greater than 30%. Overall, stronger the sequence similarity of a HP to other functional proteins, the likelihood of predicting its structure and function increases tremendously.

Computational tools have allowed researchers to generate more information about HPs in sequenced genomes across various organisms such as, mammals. Hypothetical proteins constitute a large portion of mammalian proteomes (Lubec et al., 2005) which can possibly reveal inferred evolutionary relationships between other mammals allowing for comparison. Therefore, a hypothetical protein was chosen at random from the sequenced genome of the Giant Panda (*Ailuropoda melanoleuca*). The HP chosen was PANDA_003700 and in an effort to gain insight into the process of determining the function of this HP, various computational analysis methods were implemented in the study. It was hypothesized that the function of the HP from the Giant Panda could be determined with the use of various genomic computational analysis methods. After subsequent analysis, the predicted function of the HP PANDA_003700 was that of an Mrt4 protein which is involved in ribosomal biogenesis.

References

1. Lubec G, Afjehi-Sadat L, Yang J, John JPP. (2005). Searching for hypothetical proteins: Theory and practice based upon original data and literature. *Progress in Neurobiology*,77(1–2), 90-127.

2. Galperin MY, Koonin EV. (2004). 'Conserved hypothetical' proteins: prioritization of targets for experimental study. *Nucleic Acids Research*, 32(18), 5452-5463.

3. Punta M, Coggill PC, Eberhardt RY, Mistry J, Tate J, Boursnell C, Pang N, et al. (2011). The Pfam protein families database. *Nucleic Acids Research,* 1-12.

4. Veretnik S, Bourne PE, Alexandrov NN, Shindyalov IN. (2004). Toward consistentassignment of structural domains in proteins. *Journal of Molecular Biology*, 339(3)647-78.

5. Mittl PRE, Grütter MG. (2001). Structural genomics: Opportunities and challenges.*Current Opinion in Chemical Biology,* 5(4), 402-8.

6. Zarembinski TI, Hung LW, Mueller-Dieckmann, HJ, Kim KK, Yokota H, Kim R, Kim SH.(1998). Structure-based assignment of the biochemical function of a hypothetical protein: A test case of structural genomics. *Proceedings of the National Academy of Sciences of the United States of America*, 95(26), 15189-15193.

SESSION

LATE BREAKING PAPERS

Chair(s)

TBA

An Ecological-Based Genetic Investigation Using Association Analysis Approach

Ray R. Hashemi[1], Azita A. Bahrami[2], Jeffery A. Young[1], and Aaron Schrey[3]

[1]Department of Computer Science, Armstrong State University, Savannah, GA, USA
[2]IT Consultation, Savannah, GA, USA
[3]Department of Biology, Armstrong State University, Savannah, GA, USA

Abstract— *A hybrid system of Association analysis and a modified decision tree was introduced to establish the associations between microsatellite alleles (genetic markers) of sampled lizards from eleven regions of the Florida Scrub Habitat and time-since-last-fire, TSLF (an ecological parameter), within the regions. Areas with the same TSLF across the habitat were identified and hybrid system extracted the local/global cores for each TSLF using the association rules of the area/habitat. The distributions of the patterns of the genetic markers over the areas (individually and collectively) were obtained from cores. The analysis of the distributions revealed that: (i) no particular genetic marker evenly distributed across the regions, (ii) there are local/global native patterns of DNA markers in majority of areas, (iii) areas with small TSLFs have small number of local native patterns, and (iv) the formation of the global native patterns gravitate toward the areas with more frequent past wild fires.*

Keywords: Genetic Investigation, Ecological Parameters, Association Analysis, Modified Decision Tree, Native and Transient DNA Markers, Local and Global DNA markers.

1- Introduction

Some biological, physical, and environmental data are collected for a number of lizards captured from and released into the Florida Scrub Habitat. This habitat is vast (approximately 14,000 sq. km) and lizards are sampled from eleven regions within the habitat. Each region is divided into up to four zones of north, south, east, and west. The total number of zones is 30. For each zone the time since the last natural fire or controlled prescribed fire (time-since-last-fire, TSLF) along with the number of past fires are also recorded. All the zones within one region that have the same TSLF make an *area*. The number of areas is twelve.

Among the collected biological data are the microsatellites alleles that are highly variable, generally neutral or nearly neutral genetic markers found throughout eukaryotic genomes. The use of TSLF as the ecological variable stems from the fact that fire creates important changes to habitat. In fact, fires are highly variable and can disrupt habitat, altering the characteristics of local populations. Thus, the fire is an important, yet heterogeneous, event which has significant implications to basic ecological genetics, wildlife management, and conservation efforts.

Our objective is to mine associations between microsatellite alleles (DNA markers) and TSLF (an ecological parameter) and try to answer the questions of whether a set of particular microsatellite alleles is: (i) evenly distributed across the sampled regions and (ii) native or transient to an area based on TSLF. Answering these questions is possible by establishing the *locales* for the microsatellite alleles in reference to TSLFs using the associations.

An association analysis of collected data may be completed by using Apriori algorithm [1], Formal Concept Analysis [2][3], or Rough Sets [4], to name a few. However, using any of these approaches may generate a numerous number of associations that are known as *association rules* and presented in form of *if-then* rules. To make the association rules more manageable the reduction of the rules is in order. The details of the reduction process are covered in Section three. The outcome is a set of *cores* and each core is a minimal set and reduced version of a given set of association rules.

The cores generated from an individual area's association rules and the entire habitat association rules are called *local core* and *global core*, respectively. Therefore, there are several local cores and only one global core. Cores are instrumental in identification of the locality of the microsatellite alleles.

The goal of this research effort is to establish the locales for the microsatellite alleles at both local and global levels. The locales determine the native and transient microsatellite alleles. The global locale also determines whether a set of particular microsatellite alleles is evenly distributed across the habitat.

The rest of the paper is organized as follows. The Previous Works is the subject of Section two. The Methodology is presented in Section three. The Empirical Results are discussed in Section four. The Conclusions and Future Research are covered in Section five.

2- Previous Works

The ecological-genetic based investigations have been reported in literature sporadically. Such investigations were conducted for different ecological parameters such as stressed environment [7], burned environment [8], air polluted environment [9], water polluted environment [10], etc. The work of Schrey et al. [8][11][12] is the closest to our investigation. They try to characterize the effect of fire on multiple species of lizard on Florida Scrub Habitat. Their methodology is a traditional population genetic-based statistical approach using genetic diversity, differentiation, and effective population size estimates. In contrast, our methodology is a hybrid system of association analysis and a modified decision tree which have their root in data mining. To the best of our knowledge such a hybrid system for characterizing the associations between microsatellites and TSLFs has not been reported in literature.

3- Methodology

To achieve our goal, the objectives are to extract: (i) the local and global associations between the microsatellite alleles and TSLFs, (ii) the global and local cores of TSLFs using the associations, and (iii) native and transient microsatellite alleles for TSLFs using the local and global cores.

Let $A = \{A_1 \ldots A_n\}$ be n areas within Florida Scrub habitat. From each area a number of lizards were sampled resulting in dataset of $R = \{R_1 \ldots R_n\}$ such that R_i is the set of records collected for the lizards sampled in area A_i. Each record in R_i represents the microsatellite allele markers of a sampled lizard, TSLF, and the number of past fires for the A_i.

We use the Apriori algorithm to perform the association analysis on R_i (for i = 1 to n) to produce local association rules. We also apply the same algorithm on R to generate the global association rules. The outcome for any dataset is a large number of association rules. We are interested only in those association rules in which the microsatellite alleles make the conditions and a TSLF makes the conclusion.

A *local support*, s, and a *local confidence level*, cf, are assigned to each local association rule defined by formula 1 and 2, respectively:

$$s = N/|R_i| \quad (1)$$

$$cf = N'/N \quad (2)$$

Where, N is the number of records of R_i in which only the conditions of the association rule are observed and N' is the number of records in R_i in which both conditions and conclusion of the association rule are observed. (When R is the dataset of interest and formulas (1) and (2) are changed accordingly, then s and cf for each association rule are referred to as *global support* and *global confidence level.*) An example of such rules is shown in Table 1. In these rules a condition is in form of "MA_i = v", where MA_i is a microsatellite allele and v is a value.

Table 1: A sample of Association rules

1. $MA_1 = a_1 \wedge MA_2 = b_1 \wedge MA_3 = c_3 \wedge MA_7 = m_2 \rightarrow TSLF = t_1$ *(s = 4% and cf = 85%)*
2. $MA_1 = a_1 \wedge MA_2 = b_1 \wedge MA_3 = c_3 \wedge MA_7 = m_1 \rightarrow TSLF = t_1$ *(s = 3% and cf = 95%)*
3. $MA_1 = a_1 \wedge MA_2 = b_1 \wedge MA_3 = c_3 \wedge MA_7 = m_3 \rightarrow TSLF = t_1$ *(s = 3% and cf = 90%)*
4. $MA_1 = a_1 \wedge MA_8 = q_3 \rightarrow TSLF = t_1$ *(s = 3% and cf = 95%)*
5. $MA_1 = a_1 \wedge MA_8 = q_3 \rightarrow TSLF = t_1$ *(s = 3% and cf = 95%)*
6. $MA_1 = a_1 \wedge MA_8 = q_3 \wedge MA_7 = m_2 \wedge MA_9 = p_2 \rightarrow TSLF = t_1$ *(s = 4% and cf = 85%)*
7. $MA_2 = b_2 \wedge MA_4 = d_1 \wedge MA_5 = f_2 \rightarrow TSLF = t_2$ *(s = 5% and cf = 90%)*
8. $MA_2 = b_1 \wedge MA_4 = d_2 \rightarrow TSLF = t_3$*(s = 5% and cf = 90%)*
9. $MA_4 = d_4 \wedge MA_9 = p_4 \rightarrow TSLF = t_4$*(s = 2% and cf = 52%)*
10. $MA_1 = a_2 \wedge MA_3 = c_3 \rightarrow TSLF = t_1$ *(s = 3% and cf = 90%)*
11. $MA_2 = b_2 \wedge MA_6 = h_2 \rightarrow TSLF = t_2$*(s = 5% and cf = 95%)*
12. $MA_5 = f_2 \wedge MA_8 = q_1 \wedge MA_7 = m_2 \wedge MA_7 = m_2 \wedge MA_6 = m_2 \rightarrow TSLF = t_5$ *(s = 2% and cf = 60%)*

The high volume of the association rules demands a reduction in the number of rules. the reduction is a two-phase process—*integration* and *extraction*. During the integration phase, the association rules are pruned, generalized, and collapsed using a reduction rule set. The extraction phase delivers a minimal set of reduced association rules for each TSLF—a core. The details of the integration and extraction phases are presented in the following two sub-sections.

3.1 Integration Phase

This phase is completed using the following *reduction rule set*. The order of reduction rules establishes the precedence among them and the first rule, rule (a), has the highest precedence. The reduction process continues until no more rules can be reduced.

a. Delete all the association rules with s < Threshold_Of_s or cf < Threshold_Of_cf.
b. All the rules with the same set of conditions are collapsed into a new rule that has the same set of conditions and a *conclusion set* that is the union of all the rules' conclusions.
c. Except for one condition, $C_i : MA_i = \bullet$, let the rest of the conditions, C, be the same for a group of association rules. Let also C_i differ from one rule to the next, within the group, only by the value of MA_i. If all possible values for MA_i are observed within the group then, C_i is removed from all the association rules of the group.
d. Consider the above rule again, if except for one possible value of MA_i, v', the rest of MA_i values are observed within the group then, $C_i : MA_i \neq v'$ replaces all old C_i in every association rules within the group

e. If the set of conditions in association rule of r_i is the superset of the conditions in association rule of r_j then, delete r_i and conclusion for rj changes into a conclusion set that is the union of conclusions for r_i and r_j.

As an example, we integrate the association rules of Table 1. Let us assume that the threshold value for s is 2% and threshold value for cf is 67%. The association rules of 9 and 12 are removed using reduction rule of (a). Let us also assume that all the possible values for MA_7 are m_1, m_2, and m_3. As a result, MA_7 is removed from association rules of 1, 2, and 3 using the reduction rule of (c) and all rules are collapsed into one new rule using the reduction rule of (b). The support and confidence level for the new rule will be recalculated. The association rules of 5 and 6 are removed using reduction rule of (e). No more rules can be reduced and the final integration of association rules of Table 1 is shown in Table 2. Due to the integration process, the integrated rule set does not include any association rule with the conclusions of TSLF=t_4 and TSLF= t_5 which ultimately the two TSLFs will not have cores.

Table 2: The integrated association rules of Table 1

1. $MA_1 = a_1 \wedge MA_2 = b_1 \wedge MA_3 = c_3 \rightarrow$TSLF= t_1 *(s = 4% and cf = 85%)*
2. $MA_1 = a_1 \wedge MA_8 = q_3 \rightarrow$TSLF= t_1 *(s = 3% and cf = 95%)*
3. $MA_2 = b_2 \wedge MA_4 = d_1 \wedge MA_5 = f_2 \rightarrow$TSLF= t_2 *(s = 5% and cf = 90%)*
4. $MA_2 = b_1 \wedge MA_4 = d_2 \rightarrow$TSLF= t_3*(s = 5% and cf = 90%)*
5. $MA_1 = a_2 \wedge MA_3 = c_3 \rightarrow$TSLF= t_1 *(s = 3% and cf = 90%)*
6. $MA_2 = b_2 \wedge MA_6 = h_2 \rightarrow$TSLF= t_2*(s = 5% and cf = 95%)*

Two rules with the same set of conditions but different conclusions are in *conflict*. In general, the conflicting rules are dismissed. However, we welcome the conflicting association rules in our investigation. The reason is that the conflicting rules partially represent the transient DNA markers. The rules (b), (c), (d), of the rule set and/or their combinations totally provide for handling the conflicting rules.

3.2 Extraction Phase

To extract the cores we present a modified version of the C4.5 algorithm. Briefly, the C4.5 algorithm [5] is able to build a decision tree out of the integrated rules. The decision tree delivers the patterns that uniquely identify each TSLF. We consider such patterns for each TSLF as the *TSLF core*. Due to the nature of data, the use of C4.5 (or for that matter, any decision tree algorithm) is impossible. To explain it further, the association rules of the integrated rule set have the following three properties:

Prop. 1: The number of conditions may be different from one association rule to the next.

Prop. 2: The set of conditions may be different from one association rule to the next regardless of the number of conditions.

Prop. 3: There are conflicting association rules in the integrated rules.

These properties cannot be handled by the algorithm C4.5.

To provide for these three properties, we present a modified version of C4.5, named MC4.5, using some of the foundations of the modified ID3 (MID3) proposed by the authors [6]. The details of the MC4.5 are given in the following sub-section.

3.2.1 MC4.5

The MC4.5 follows the same principals of the C4.5 algorithm which generates a decision tree out of the integrated rule set extracted from R denoted by P_R. That is, MC4.5 also uses the *gain ratio* to determine the root for the decision tree and every sub-decision tree. Considering the fact that R={R_1 . . . R_n}, there are n unique values among the conclusion values (K) in P_R. Let $\mu(P_R)$ be defined as;

$$\mu(P_R) = -\sum_{i=1}^{n}\left({f_{K_i}}/{|P_R|}\right)Log_2\left({f_{K_i}}/{|P_R|}\right) \quad (3)$$

Where $f_{K_i}$is the frequency of conclusion K_i in P_R.

A condition of C_i may have m possible values. Selecting C_i as the root of the decision tree (or a sub-tree) generates m branches and each branch is a partition of P_R, P_R^j, for j = 1 to m . The *gain ratio* for C_i is calculated using formula 4.

$$\text{Gain Ratio}(C_i) = \frac{\gamma(C_i)}{\lambda(C_i)} \quad (4)$$

Where,

$$\gamma(C_i) = \mu(P_R) - \sum_{j=1}^{m}\mu(P_R^j)\left({|P_R^j|}/{|P_R|}\right) \quad (5)$$

and

$$\lambda(C_i) = -\sum_{i-1}^{n}(|P_R^j|)Log_2(|P_R^j|) \quad (6)$$

The condition with the highest gain ratio is the winner of the competition for serving as root of the decision tree or root of a sub-tree. The same process is repeated for each child of the winner until all the conclusions for the association rules in a given leaf of the tree are the same.

To provide for the properties of 1 and 2, MC4.5 will adopt the following procedure. Let us concentrate one more time on the condition of C_i with m possible values. To evaluate C_i for serving as the root of the tree or a sub-tree the number of branches for C_i is always m+1. The extra branch for C_i includes all the association rules in which C_i does not exist. The branch is labeled *null*.

To handle the last property, MC4.5 will adopt the following procedure. Let us assume that there is one rule with a conclusion set among the association rules that belong to a given leaf of the tree. This means that the association rules of the leaf has to be partitioned further using another condition as the root of the new subtree. Let us also assume that the number of unique conclusions for the entire association rules of the leaf is n'. To handle the conclusion set (conflicting rules) we use one of the TSFL values in the conclusion set as the favorite conclusion for the association rule and mark the rule with indices such that in future one can remember that other TSLF values are also related to the rule. The favorite conclusion is chosen in such a way that n' is reduced.

After the decision tree is built, each path of the decision tree makes an entry into the core of the TSLF, a *corlet*, for the leaf of the path in form of an if-then rule. In addition, each marked corlet is duplicated as many times that it has been marked and the conclusion for each copy is one of the TSLF values in the conclusion set. The duplicated corlets are added to their related cores.

As an example, a set of association rules that also includes conflicting rules are given in Table 3. (For simplicity, the support and confidence level of the rules are omitted.) Rules 1 and 6 replaces by: $MA_1 = a_1 \wedge MA_2 = b_1 \wedge MA_9 = u_1 \rightarrow TSLF= \{t_1, t_2\}$ using reduction rule b. Using the same reduction rule replaces rules 7 and 8 by: $MA_1 = a_1 \wedge MA_2 = b_1 \rightarrow TSLF= \{t1, t_3\}$. The new two rules are replaced by $MA_1 = a_1 \wedge MA_2 = b_1 \rightarrow TSLF= \{t_1, t_2, t_3\}$ using reduction rule (c). Rules 2 and 10 are replaced by rule: $MA_1 = a_1 \wedge MA_8 = q_3 \rightarrow TSLF= \{t_1, t_6\}$. The integrated dataset (the global core) is shown in Table 4.

Table 3: A given dataset with conflicting rules

1.	$MA_1 = a_1 \wedge MA_2 = b_1 \wedge MA_9 = u_1 \rightarrow TSLF= t_1$
2.	$MA_1 = a_1 \wedge MA_8 = q_3 \rightarrow TSLF= t_1$
3.	$MA_2 = b_2 \wedge MA_4 = d_1 \wedge MA_5 = f_2 \rightarrow TSLF= t_2$
4.	$MA_2 = b_1 \wedge MA_4 = d_2 \rightarrow TSLF= t_3$
5.	$MA_1 = a_2 \wedge MA_3 = c_3 \rightarrow TSLF= t_1$
6.	$MA_1 = a_1 \wedge MA_2 = b_1 \wedge MA_9 = u_1 \rightarrow TSLF= t_2$
7.	$MA_1 = a_1 \wedge MA_2 = b_1 \rightarrow TSLF= t_1$
8.	$MA_1 = a_1 \wedge MA_2 = b_1 \rightarrow TSLF= t_3$
9.	$MA_2 = b_2 \wedge MA_6 = h_2 \rightarrow TSLF= t_2$
10.	$MA_1 = a_1 \wedge MA_8 = q_3 \rightarrow TSLF= t_6$

Table 4: The result of applying integration phase on Table 3

1.	$MA_1 = a_1 \wedge MA_2 = b_1 \rightarrow TSLF= \{t_1, t_2, t_3\}$
2.	$MA_1 = a_1 \wedge MA_8 = q_3 \rightarrow TSLF= \{t_1, t_6\}$
3.	$MA_2 = b_2 \wedge MA_4 = d_1 \wedge MA_5 = f_2 \rightarrow TSLF= t_2$
4.	$MA_2 = b_1 \wedge MA_4 = d_2 \rightarrow TSLF= t_3$
5.	$MA_1 = a_2 \wedge MA_3 = c_3 \rightarrow TSLF= t_1$
6.	$MA_2 = b_2 \wedge MA_6 = h_2 \rightarrow TSLF= t_2$

The MC4.5 generates the decision tree of Figure 1 for the integrated rules of Table 4. The corlets for the core of each TSLF is shown in Table 5.

3.3 Core Analysis

It is significant to express that the local core for $TSLF_i$ is the same as the integrated rules of $TSLF_i$. The reason stems from the fact that all the integrated rules for $TSLF_i$ have the same conclusions. And application of MC4.5 does not reduce the integrated rules further.

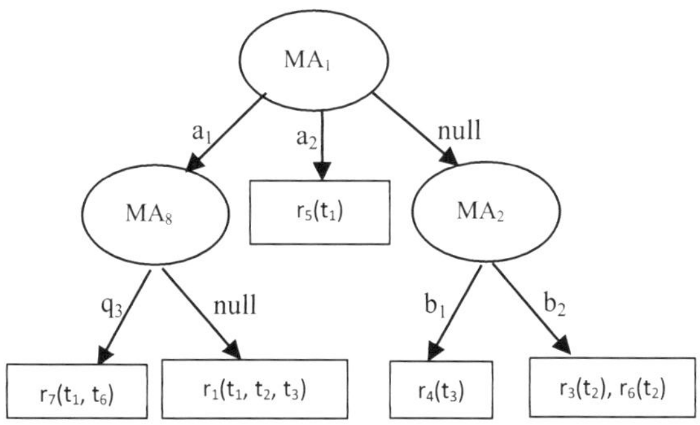

Figure 1: The decision tree for the integrated rules of Table 4 generated by MC4.5

Table 5: The global cores for different TSLFs

TSLF	Global Cores
t_1	$MA_1 = a_1 \wedge MA_8 = null \rightarrow TSLF= t1$ $MA_1 = a_2 \rightarrow TSLF= t_1$ $MA_1 = a_1 \wedge MA_8 = q_3 \rightarrow TSLF= t_1$
t_2	$MA_1 = null \wedge MA_2 = b_2 \rightarrow TSLF= t_2$ $MA_1 = a_1 \wedge MA_8 = null \rightarrow TSLF= t_2$
t_3	$MA_1 = null \wedge MA_2 = b_1 \rightarrow TSLF= t_3$ $MA_1 = a_1 \wedge MA_8 = null \rightarrow TSLF= t_3$
t_6	$MA_1 = a_1 \wedge MA_8 = q_3 \rightarrow TSLF= t_6$ $MA_1 = null \wedge MA_2 = b_2 \rightarrow TSLF= t_6$

A core satisfies the following criteria:

1. Core is free of redundant association rules,
2. Core has minimum cardinality,
3. The number of conditions for any given association rule in core cannot be reduced further,
4. Core preserves the associations of its superset, and
5. There is only one core per a given set of associations.

The purpose of core analysis is to identify the transient and native DNA markers (locales) at both local and global levels. The transient markers are identified using *Core_Analysis* algorithm that finds the common patterns (i.e. conditions) between any two given association rules of r_i and r_j that belong to two different cores. Two cores belong to two different TSLFs' areas separated from each other with $dist(A_i, A_j) < nh_r$, where nh_r is the neighborhood radius. The nh_r is set to include only the immediate neighbors of a given area and set to infinity when the algorithm is applied on the local cores and global cores of the area, respectively. The algorithm also determines the support(s) and confidence level (cf) for the common pattern and dismisses those common patterns for which either the support or the cf is less than a threshold. These thresholds may be different from the thresholds of s and cf used in the reduction rule set.

Algorithm Core_Analysis

Given: Core of area, core(A_i). Corlet$_j$(A_i) that refers to the j-th association rule in core(A_i). A file named Sink with the record layout of <Area 1> <Area 2> <common pattern> <count>. The neighborhood radius of nh_r. The threshold T_s for support and the threshold T_{cf} for confidence level

Objective: Identify the transient patterns of microsatellite alleles among the TSLFs.

1: Repeat for k = 1 to n-1
2:   g = k+1;
3:     Repeat for i = 1 to $|core(A_k)|$
4:       Repeat for j = 1 to $|core(A_g)|$
          If (dist(A_k, A_g)) > nh_r Then break;
5:         $\Phi = corlet_i(A_n) \cap corlet_j(A_g)$;//common pattern
6:         m_1= the number of records in core(A_k) in which $\Phi$is observed;
7:         m_2= the number of records in core(A_g) in which $\Phi$is observed;
8:         $s_\Phi = (m_1 + m_2)/(|core(A_k)| + |core(A_g)|)$;
9:         cf_Φ = Min(corlet$_i$(A_n).cf, corlet$_j$(A_g). cf);
10:        If ($\Phi \neq \emptyset$ && $s_\Phi > T_s$ && $cf_\Phi > T_{cf}$)
          Then P $\leftarrow$ concatenate (A_K, A_g, Φ);
            If (P matches a record in Sink)
            Then increment the count of the record;
            Else append P to Sink with count = 1,
11:      End Repeat;
12: g ++;
13:   End Repeat;
14: End Repeat;
End;

Each record of the file Sink is displayed by a code copied in both areas of <Area 1> and <Area 2>. The code is made-up of a character followed by a number. The character and the number represent the transient microsatellite pattern (Φ), and *strength* of the pattern (<count>), respectively.

Now, we present the identification of the local and global locales. Let the core for the area (A_i) with TSLF$_i$ be P_i and let also the areas with TSLF$_1$ . . . TSLF$_m$ be the immediate neighbors of A_i with the cores of Q_1 . . . Q_m. The local transient patterns (LTP) and the local native patterns (LNP) to the TSLF$_i$, are defined using formula (7) and (8).

$$\text{LTP}(\text{TSLF}_i) = \cup_{j=1}^{m} (P_i \cap Q_j) \quad (7)$$

$$\text{LNP}(\text{TSLF}_i) = P_i - \text{LTP}(\text{TSLF}_i) \quad (8)$$

The local native patterns of an area refer to the patterns that are unique to the area in reference to the immediate neighboring areas. And they signify the local locales.

As an example, consider the integrated association rules of Tale 4. Let us assume that the area with TSL = t_2 is the only immediate neighbor of TSLF=t_1 area. The local transient patterns are: $MA_1 = a_1(2)$ and $MA_2 = b_1(1)$. The local native patterns (i.e. the local locales) are: $MA_8 = q_3$, $MA_2 = a_2$, and $MA_3 = c_3$.

Let the global core of the TSLF$_j$ be G_j. The global transient patterns for TSLFi, GTP(TSLF$_i$), are identified using formula (9).

$$\text{GTP}(\text{TSLF}_i) = \cup_{j=1}^{m} \left(G_i \cap G_j\right), (i \neq j) \quad (9)$$

The *global native patterns* (global locales) for the TSLF$_i$, GNP(TSLF$_i$), are identified using formula (10).

$$\text{GNP(Ai)} = G_i - GTP(TSLF_i) \quad (10)$$

The global transient patterns and the global native patterns for the cores of Table 5 are shown in Table 6.

Table 6: Global transient and native patterns to TSLFs

TSLF	Transient patterns	Global Locales to TSLF
t1	$MA_1 = a_1$ (2), $MA_2 = b_1(1)$, $MA_8 = q_3$ (1)	$MA_2 = a_2$ $MA_3 = c_3$
t2	$MA_1 = a_1$ (2), $MA_2 = b_1(1)$	$MA_2 = b_2$, $MA_2 = d_1$, $MA_5 = f_2$, and $MA_2 = h_2$
t3	$MA_1 = a_1$ (2), $MA_8 = q_3(1)$	$MA_4 = d_2$
t6	$MA_1 = a_1$ (2), $MA_8 = q_3(1)$	None

4- Empirical Results

Data collected by sampling 429 lizards in Florida Scrub Habitat. The sampled lizard selected from eleven regions and 12 areas within the regions. The record of each lizard was made up of eight DNA markers (microsatellite alleles), area name from which the lizard was sampled, time-since-last-fire (TSLF) for the area, and the number of the past fires for the area. After removing the outliers from the dataset, we had 352 records.

The Apriori algorithm used to generate the local and global association rules separately. The local cores and global core were extracted. The cores were analyzed using the Core_Analysis algorithm. The results for the local transient patterns are shown in Figure 2 that is a schematic representation of the Florida Scrub Habitat.

Each area, in Figure 2, is shown as a rectangle with a small rectangle shape tag that carries the name of the area followed by two pairs of parentheses. The first pair shows the area's TSLF in bold and the second pair shows the number of past fires for the area in bold. Due to space limitations in Figure 2, each pattern is shown by its code. The meaning of each character used in pattern codes of Figure 2 is given in Table 7.

MC4.5 algorithm was used to generate the global cores for each TSLF. The global transient DNA patterns were extracted and shown in Figure 3. Areas across the habitat with the same TSLF have the same border patterns that are not solid lines. The meaning of each character used in pattern codes of Figure 3 is also shown in Table 7. The local and global native DNA patterns extracted by applying the core analysis are shown in Table 8 and Table 9, respectively.

Table 7: Meaning of characters used in Figure 2 and Figure 3

Pattern & Code		Pattern & Code		Pattern & Code	
Nr52.7=23	a	Nr52.11=41	b	Nr52.11=44	c
Nr52.2=40	d	Nr52.2=42	e	Nr52.2=43	f
Nr52.4=20	g	Nr52.4=25	h	Nr52.4=26	i
Nr52.4=27	j	Nr52.7=17	k	Nr52.7=19	l
Nr52.7=29	m	Nr60.11=14	n	Nr60.11=17	o
Nr60.2=16	p	Nr60.2=21	q	Nr60.34=10	r
Nr60.34=11	s	Nr60.34=20	t	Nr60.5=12	u
Nr60.5=13	v	Nr 52.4=24	w	Nr60.11=12	x

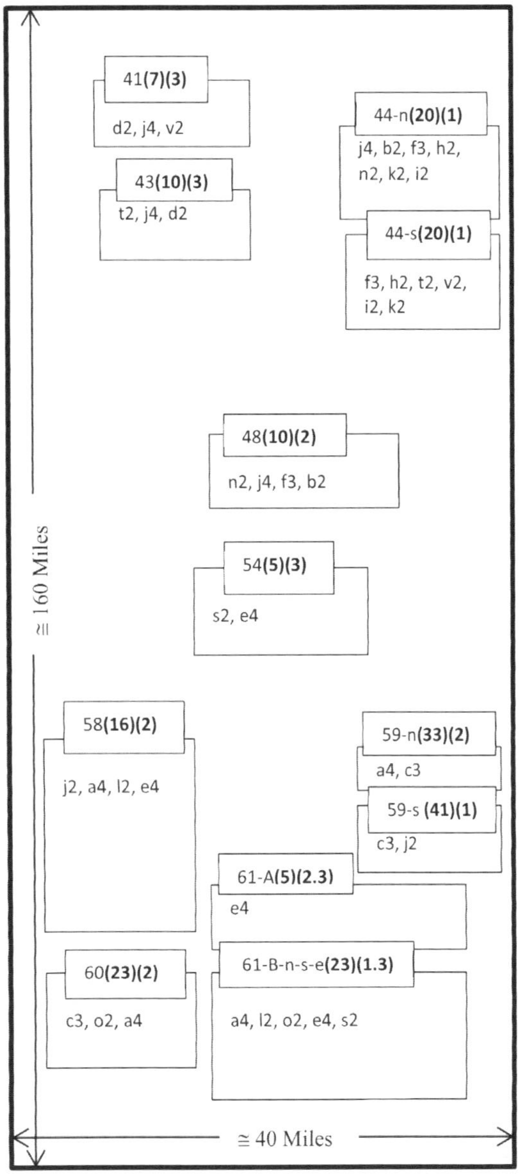

Figure 2: Local transient patterns for the Scrub habitat.

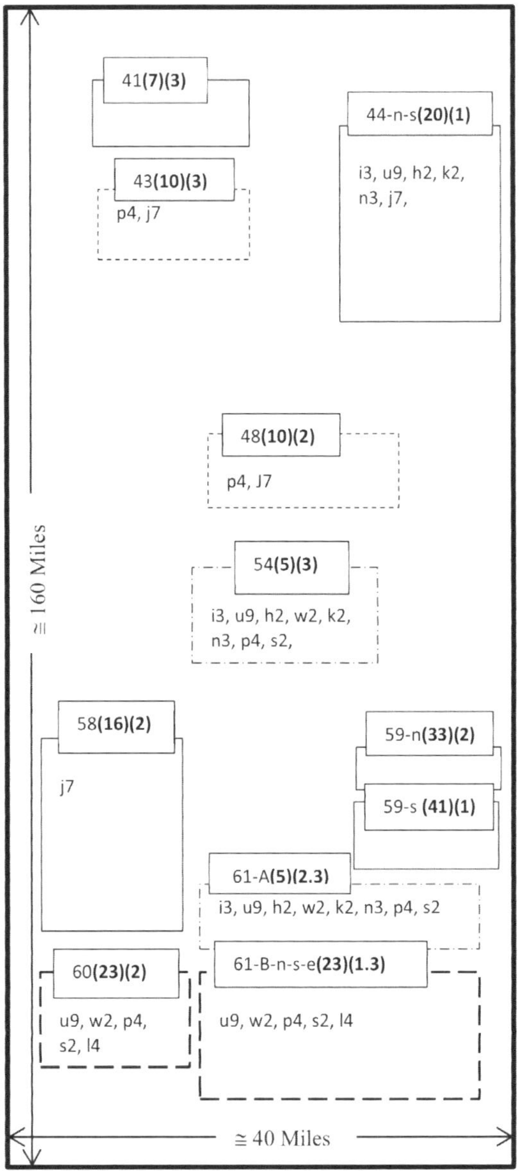

Figure 3: Global transient patterns for the Scrub habitat.

5- Conclusions and Future Research

Both Figures of 2 and 3 revealed that there is not any particular set of microsatellite allele that is evenly distributed across the sampled areas. Some of the microsatellites do not have any locale due to the fact that the association rules for those microsatellites have been deleted during the integration phase. The largest group of local native DNA markers belongs to 61-B and 44-n with TSLFs of 23 and 20 years, respectively. And they also have the largest group of transient DNA markers. Both areas are located at the extremes of north-east and south-east of the habitat and more importantly the number of fires in the past for both areas is equal to one (which means minimum.)

In general, when the TSLF is small, the number of local native patterns is also small. This finding supports the results from a traditional population genetic statistical approach where the most recent fires were shown to have a disruptive effect on genetic characteristics [8].

The common DNA markers among the local and global transients for the similar areas are: Nr52.4=27, Nr52.4=26, Nr52.4=25, Nr52.7=17, Nr52.7=19, Nr60.34=11, and Nr60.11=14. That is, all the lizards with these DNA markers are *super hyperactive*.

Table 8: Locales (native patterns) driven from local cores

TSLF in Years and (Area Name)	Locale (Local Native Pattern)	TSLF in Years and (Area Name)	Locale (Local Native Pattern)
7 (41)	Nr60.5=12	16 (58)	None
10 (43)	Nr52.7=23 Nr52.11=44 Nr60.2=21	33 (59-n)	None
20 (44-n)	Nr52.7=19 Nr52.7=29 Nr60.34=10	41 (59-s)	Nr52.4=20
25 (44-S)	None	23 (60)	Nr52.7=17
10 (48)	Nr60.2=16	5 (61-A)	None
5 (54)	Nr52.4=26 Nr52.7=29	23 (61-B)	Nr52.2=43 Nr52.4=25 Nr52.11=41 Nr60.2=16 Nr60.34=10 Nr60.5=12 Nr60.5=13

Table 9: Locales (native patterns) driven from global cores.

TSLF in Years and (Area Name)	Locale (Global Native Pattern)	TSLF in Years and (Area Name)	Locale (Global Native Pattern)
5 (54, 61A)	Nr52.2=41 Nr52.7=29 Nr60.2=17 Nr52.11=29 Nr60.11=12	10 (43, 48)	Nr60.34=20 Nr52.2=40 Nr60.2=21 Nr60.2=19 Nr52.7=15 Nr52.7=44 Nr60.5=14
7 (41)	None	16 (58)	Nr52.7=23 Nr60.34=29
20 (44)	Nr52.4=21 Nr60.5=17 Nr60.2=22	23(60, 61B-n-s-e)	Nr52.2=36 Nr52.4=20 Nr52.11=41 Nr60.2=18 Nr60.5=10
33(59-n)	Nr52.2=45		

The native patterns that remain with the areas in both local and global levels are *consistent native patterns* and they are Nr60.2=21 in area (43, 48), Nr52.11= 41in area (60, 61-B), and Nr52.7=29 in area (54). The average numbers of past fires for these areas are: 2.5, 2, and 3, respectively. It is interesting that the consistent native patterns are related to the areas with the high number of past fires. That is, all the lizards with these DNA markers are die hard natives. We have also observed that the formation of global native patterns gravitates toward the areas with large number of past fires.

As future research, the investigation of the relationships between native patterns (local and global) and DNA diversity is in progress.

References

[1] Hashemi, R. R., L. LeBlanc L., Westgeest, B., "The Effects of Business Rules on the Transactional Association Analysis", The 2004 International Conference on Information Technology: Coding and Computing (ITCC-2004), Pradip K. Srimani (Editor), Sponsored by IEEE, Las Vegas, Nevada, April 2004, Vol. II, pp. 198 - 202.

[2] Hashemi, R.R., Le Blanc, L., Bahar, M., and Traywick B., "Association Analysis of the Alumni Data Using Formal Concept Analysis", the 15th International Workshops on Conceptual Structures (ICCS'07), Babak Akhgar (Editor) Springer-Verlag Publisher, Sheffield, UK, July 2007, pp. 187-193.

[3] B. Ganter, and R. Wille (1999), Formal Concept Analysis: Mathematical Foundations, Berlin: Springer-Verlag.

[4] Hashemi, R. R., Tyler, A, Bahrami, A. "Use of Rough Sets as a Data Mining Tool for Experimental Bio-Data", A book chapter in: "Computational Intelligence in Biomedicine and Bioinformatics: Current Trends and Applications", Tomasz G. Smolinski, Mariofanna G. Milanova, and Aboul Ella Hassanien, Editors, Springer-Verlag Publisher, June 2008, pp. 69-91.

[5] Quinlan, J. R., "C4.5: Programs for Machine Learning", Morgan Kaufmann, 1993.

[6] Hashemi, R.R., Le Blanc, L., Bahar, M., and Traywick B., "Profiling the Potential Donors to a Non-Profit Organization", The 2007 International Conference on Information and Knowledge Engineering (IKE'07), Las Vegas, Nevada, June 2007, pp. 343-348

[7] Geburek TH. and Knowles P., "Ecological-genetic Investigations in Environmentally Stressed Mature Sugar Maple (Acer Saccharum Marsh) Populations, Water, Air, and Soil Pollution, Vol. 62, 3: 261-268, 1992.

[8] Schrey A., Heath S., Ashton K., McCoy E., and Mushinsky H., "Fire alters patterns of genetic diversity among three lizard species in Florida scrub habitat, Journal of Heredity, 102: 399-408, 2011.

[9] Taylor, Jr., G. F., Pitelka L.F., (Eds.), *"Ecological Genetic and Air Pollution"*, Springer-Verlog Publisher, 1991.

[10] Bert T. M. (Ed.) *"Ecological and Genetic Implications of Aquaculture Activities"*, Springer-Verlog publisher, 2007.

[11] Schrey A., Fox A., McCoy E., and H. Mushinsky "Fire increases variance in genetic characteristics of Florida Sand Skink (*Plestiodon reynoldsi*) local populations. Molecular Ecology. 20: 56-66, 2011.

[12] Schrey A., Ragsdale AK., McCoy E., and Mushinsky H., "Fire-based habitat disturbances can decrease effective population size of local populations. Journal of Heredity. 107:336-341, 2016.

Author Index

AlEisa, Hussah - 50
Armaselu, Bogdan - 67
Arunachalam, Harish - 67
Ashkani, Johanshah - 109
Bach, John-Paul - 67
Bahrami, Azita A. - 135
Bixby, Cameron - 130
Bridges, Susan M. - 44
Bullock, Caroline - 3
Cai, Hong - 97
Cederberg, Kevin - 67
Chai, Lian En - 50
Chan, Weng Howe - 50
Che, Dongsheng - 31
Chen, Bernard - 31
Cogill, Steven - 38
Daescu, Ovidiu - 67
Deller, E. John R. - 81
Deris, Safaai - 50
Eyre, Jane - 117
Garcia-Nocetti, F. - 90
Glick, Sam - 67
Goodman, Erik D. - 81
Hamel, Lutz - 10
Hardy, Christopher - 129
Hardy, Nazli - 129
Hashemi, Ray R. - 135
Jaegle, Stephen - 10
Jain, Chakresh - 56
Jiang, Jiayao - 85
Jung, Hyedong - 103
Kang, Sang-Ick - 127
Kim, Bongjae - 103
Kim, Sang-Kyun - 127
Kwon, Jang-Woo - 127
Leavey, Patrick - 67
Lee, Hyungdo - 103
Lee, Kyounghak - 103
Lee, Sangmin - 127
Li, Luxi - 81
Li, Wing Ning - 17
Li, Zuqing - 31
Lilburn, Timothy - 97
Lim, Seonjoo - 10
Mahadevan, Padmanabhan - 130
Manda, Prashanti - 44
Manikonda, Sai - 31
McCarthy, Fiona - 44

McFarland, Devaun - 3
Mohamad, Mohd Saberi - 50
Mueller, Benjamin - 3
Mustard, Julie - 22
Nanduri, Bindu - 44
Ngo, Jessica - 109
Park, Young-Jin - 127
Passi, Kalpdrum - 56
Pio, Frederic - 109
Rakheja, Dinesh - 67
Reader, Sara - 22
Richfield, Steve - 117
Rubio, E. - 90
Schrey, Aaron - 135
Seidler, Norbert - 22
Sengupta, Anita - 67
Shah, Zuraini Ali - 50
Shi, Fengxing - 85
Sirrianni, Joseph - 17
Skapek, Stephen - 67
Solano, J. - 90
Subair, Saad - 50
Sun, Jing - 56
Valafar, Homayoun - 3
Wang, Hui - 44
Wang, Liangjiang - 38
Wang, Yufeng - 97
Warner, Robert - 74
Xiao, Zhichun - 17
Yao, Meng - 81
Yao, Yizhou - 81
Young, Jeffery A. - 135
Zainuddin, Muhammad Mahfuz - 50
Zhang, Chunkai - 85
Zhang, Hao - 97